MW01601116

UFO Sightings of 2006-2009

Scott C. Waring

iUniverse, Inc.
New York   Bloomington

iUniverse books may be ordered through booksellers or by contacting:

iUniverse
1663 Liberty Drive
Bloomington, IN 47403
www.iuniverse.com
1-800-Authors (1-800-288-4677)

ISBN: 978-1-4502-3239-5 (sc)
ISBN: 978-1-4502-3240-1 (hc)
ISBN: 978-1-4502-3241-8 (ebook)

Library of Congress Control Number: 2010907421

Printed in the United States of America

iUniverse rev. date: 05/13/2010

"I occasionally think how quickly our differences worldwide would vanish if we were facing an alien threat from outside this world."

President Ronald Reagan speaking at the United Nations.

Forward

My goal is not to flood you with unsubstantiated information or stories. Instead my goal is to raise humanities level of awareness of the species around us.

Many debunkers of today are hired by the CIA and NSA, to try to stop any leaked information about UFO's and make it appear that the individual who released the information is mentally unbalanced. They try to base their efforts upon a single psychological term called pareidolia, which is when a person notices an illusion or misperception involving a vague and obscure stimulus such as an image or sound as being significant. Much as anyone may look into the clouds and see shapes like animals, faces and such. Pareidolia is a bluffing technique of which the U.S. Government will try to use when faced with a real UFO phenomenon, but they only pull this card when genuine UFO information leaked has actually succeeded in trickling down to Joe Public.

It was Sigmund Freud who once said, "Sometimes a cigar is just a cigar." Meaning here that sometimes a UFO sighting is just a UFO sighting and not something less. Often people try to justify something as being ones imagination by reading too much into something that is relatively simplistic in nature. This is what the government depends upon; to make you Joe citizen, believe that the person trying to tell you the truth is actually crazy. This is known as 'herd poisoning' and was perfected by Hitler himself to control the thoughts and actions of his countrymen. It is a last desperate effort to control the uncontrollable…the truth.

It was Psychologist Carl Rogers who said, "The truth will set you free, but first it will hurt like hell." I found those words to be true wisdom when I uncovered the reality of the alien presence throughout our universe by using only NASA photographs. Cameras do not lie unlike humans, and using my simple technique to make those photos larger and 100% clear, you too will become one of the few who knows the truth about the existence of aliens. So when you see someone spending many hours trying to debunk a UFOlogist's work, stop and ask yourself, "Why are they working so hard for free, just to make the UFO evidence appear as a fraud?" Perhaps the person is a government employee and it is his/her job to stop those leaks ASAP. The Internet is their tool. When debunkers begin focusing on one UFOlogist, when a UFOlogist tells the world their story and then suddenly dies, or has to stay in hiding, or is hounded by the government, that's where I come in to research them and see if their story is true and what evidence that they have to support their allegations. If evidence says they are telling the truth, then I will dive more deeply to see if I can carry on where they left off.

Chapter One: O'Hare Airport UFO of 2006

One of the most talked about sightings in 2006 came on November 7th at the second busiest airport in America, the O'Hare. A saucer like object hovered over the airport for between three and seven minutes, before it shot straight upward at tremendous speeds. It shot through the clouds that sat above it, causing a circular gapping hole to be made as the saucer passed through it and disappeared. The circular hole in the cloud remained opened for fifteen minutes before it too disappeared. Although many UFOlogist call these flying objects Unidentified Aerial Phenomenon, I believe political correctness or extreme accuracy of naming the craft is less significant than proving the actual existence of the craft, so I will here forth refer to them as UFO's.

1. Photo of O'Hare UFO over control tower.
   (At http://scwbook.blogspot.com/)

2. Photo of O'Hare UFO over runway.
   (At http://scwbook.blogspot.com/)

Associated Press wrote an article about O'Hare in July 3, 2006 quoting government statistics. It said O'Hare was the busiest airport in the nation for the first half of 2006. The airport had 477,001 flights including both take-offs and landings in that first six months. O'Hare's air traffic controllers usually have close to ninety-six arrivals per hour. That is one every 38 seconds often on numerous runways.

In the beginning the officials at United Airlines said that they had no information about the sighting of the UFO. The Chicago Tribune came to inquire about the sighting and had over one dozen United Airlines employees report to them that they witnessed the event first hand. The Federal Aviation Administration however did say that they received one phone call from a United Airlines supervisor about some of the employees seeing an unusual object that was elliptical shaped hovering motionless over Concourse C United terminal.

None of the controllers at O'Hare airport had seen the UFO and a check of radar activity revealed that nothing out of the ordinary was anywhere in the area. If this is true and current radar is incapable of picking up or tracking UFO phenomenon, then the government should look into improving such radar technologies, unless of course the UFO was seen on radar and they merely want to deny its existence, they will just deny and discredit all witness accounts of the event, saying it was a weather balloon or weather occurrence. Both are typical excuses used by the government to cover up sightings that are deemed classified material. An unknown object said to be fifteen to twenty five feet in diameter was hovering over an active runway, with the US already on high alert after the 911 attacks, yet they end the investigation as fast as it started. Not to mention that it posed a serious safety risk to the pilots and the passengers.

The FAA announced that they would not continue to investigate into the matter further. This news upset many who witnessed the sightings. This kind of decision shows the cover up in progress. Many of the employees interviewed by the Chicago Tribune stated that they were extremely upset that the government and United Airlines stopped all additional examination into the sighting. One such witness stated, "Some of us are getting angry with this being hushed up with all the terrorism and TSA idiots hanging around. If we see a funny looking bag all damn hell breaks loose, but park a funny silver thing a few hundred feet above a busy airport and everyone tries to hush it up. It just doesn't make sense."

Craig Burzych, a supervisor for United laughed about the whole ordeal saying that, "to fly 7 million light years to O'Hare and then to have to turn around and go home because your gate was occupied is simply unacceptable." This kind of humor about such a serious incident is actually quite commonplace and makes the whole event look as if were a minor matter. He wasn't the only one to laugh about the event. There were many conversations between pilots and mechanics who witnessed the UFO firsthand and some pilots who where just curious joking to the control tower latter that day.

The first person to see the flying object was a United ramp worker who was directing back a United plane at Gate C17. He saw the UFO at 4:15 p.m. just as the sun began to set in the distance. Note that many sightings do occur during sunset, meaning that perhaps the changing light influences the UFO's translucent camouflage appearance making it visible momentarily.

A few of the witnesses said that the ship appeared like a rotating Frisbee in the sky. Other witnesses said the UFO was not appearing to rotate. Even though they differ in opinions about the rotation, they all agreed that the flying disk made no noise whatsoever and that it was situated in a motionless position just below the 1,900-foot cloud deck, before it shot upwards at a slight angle. The size of the UFO seems to vary among eyewitnesses at being between twenty-five and eighty-eight feet in diameter.

A mechanic who refused to give his name, yet worked for United Airlines was clearly baffled by the whole situation. "I tend to be scientific by nature, and I don't understand why aliens would hover over a busy airport." He was the earliest known witness at the airport. He was in the cockpit of a 777 Boeing when he witnessed the UFO in the sky. He was standing on the tarmac beside the nose of the 777 when he was compelled to look straight up for some reason and was surprised to see the UFO hovering. He watched the UFO as he taxied the 777 into its hanger. Perhaps he is right about the situation. It might not have been a UFO, but a cocky USAF pilot thinking his alien technology was further advanced than it actually was. We have to be open-minded when it comes to the unknown. The mechanic continued, "but I know that what I saw and what a lot of other people saw stood out very clearly, and it definitely was not an Earth aircraft." The fact that the mechanic said he was compelled to look up for some unknown reason is similar to other sightings (for instance Seattle) where people were compelled to get out of bed at three in the morning and walk outside to see unusual lights in the sky. This insinuates that some crafts have a telepathic connection that can reach interspecies levels, assuming the UFO was alien controlled.

One employee who had witnessed the UFO was rumored to be having some serious issues dealing with her religion after the event. This too seems logical since one of the things that was the deciding NASA questions for early Apollo astronauts was, "Do you believe in God?" If the answer was no, then you had the right stuff, or so a supposedly former astronaut William Rutledge said in an interview. Perhaps fear of classified information leaking if an astronaut seeks a session of religious counseling could cause NASA to be concerned.

A supervisor at United airlines said that he ran outside his office at Concourse B right after hearing reports of the UFO from other employees talking about it over the airline radio frequencies. "I stood outside in the gate area not knowing what to think, just trying to figure out what it was." He said he knew that employees would not make such a false report as this, so he had to see it for himself. He felt that it might be a weather balloon or something else hovering over O'Hare and may have to be stopped before it came into close proximity to the flights.

In a transcript of the conversation that the United Airlines ramp tower and the FAA Area Supervisor in a near by O'Hare tower, we begin to understand some details through their interesting and revealing conversation that started at 4:30 p.m.

> O'Hare Tower: "Tower, this is Dave."
>
> Ramp Tower: "Hey Dave, this is Sue in the United Tower."
>
> O'Hare Tower: "Hey Sue."
>
> Ramp Tower: "Hey, did you see a flying disc out by C17?"
>
> O'Hare Tower: "Oh, it starts Sue." Then they laughed. "Oh, we're sorry Sue. A flying...you're seeing flying discs?"
>
> Ramp Tower: "Well, that's what a pilot in the ramp area at C17 told us. They saw some flying disc above them. But we can't see above us."
>
> O'Hare Tower: "Common Sue."
>
> Ramp Tower: "You didn't see it?"
>
> O'Hare Tower: "No," she laughs.
>
> Ramp Tower: "Hey, you guys been celebrating the holidays or anything, or what? You're celebrating Christmas day? I haven't seen anything Sue, and if I did I wouldn't admit to it. No I have not seen any flying disc at gate C17." Here Sue continues laughing. "Unless you've got a new aircraft you're bringing out that I don't know about."
>
> O'Hare Tower: "No," she laughs.

This conversation shows how even an experienced airport employees have fears about discussing information about UFO sightings. The fears that the two individuals have about discussing such a topic causes them to laugh when they try to imagine it. This is the kind of attitude that the US Government is counting on in order to keep the details they have learned to themselves. The government uses psychology to play on Joe Publics' fears of looking foolish, and it works almost every time. In their next conversation the fact that a picture does actually exists gets uncovered.

> Ramp Tower: "I'm sorry, there was, I told Dave, there was a disc flying outside above Charley 17 and he thought I was pretty much high. But, um, I'm not high and I'm not drinking."
>
> O'Hare Tower: "Yeah."
>
> Ramp Tower: "So, someone got a picture of it. So if you guys see it out there…"
>
> O'Hare Tower: "A disc, like a Frisbee?"
>
> Ramp Tower: Like a UFO type thing."
>
> O'Hare Tower: "Yeah, Okay."
>
> Ramp Tower: "He got a picture of it."
> (Laughing)

A third conversation was recorded between a male at United Ramp tower and O'Hare tower that began at 4:52 pm. It admits to multiple pilots witnessing the UFO.

> Ramp Tower: "Some of our employees…I don't know if you know anything about this? Some of our pilots on the ground are reporting a UFO sighting at a thousand feet to the east side of

the airport. Do you guys know anything about this?"

O'Hare Tower: "You know, the ramp tower called me, I want to say about ten or fifteen minutes ago. We have not seen anything up here."

Richard Haines, who is the science director at the National Aviation Reporting Center on Anomalous Phenomena, was deeply concerned with the sighting. "More and more we are coming to the point of view that we are dealing with an intelligent phenomenon." Haines reporting center is a privately run agency. He feels that people need to be proactive now before an aircraft accident occurs. Haines is also a former chief of Space Human Factors Office at NASA's Ames Research Center.

All the people who saw the O'Hare UFO firsthand, including a few pilots, said they are confident that based upon the overall appearance of the strange craft, that there was no way possible that it could be an airplane, weather balloon, helicopter or any other craft that is known to exist. The witnesses were unsure of what controlled the craft that they saw, yet very certain that it was saucer shaped. The employees at United were irritated that a UFO came into restricted airspace for a few minutes and no one in power had taken the matter seriously. This seems to be an opposite behavior than we would expect from experts at the FAA and O'Hare, so maybe word from above was sent down through the chain of command to hush it ASAP before a media frenzy began.

In the beginning, United Spokeswoman Megan McCarthy said that there was no record of any UFO report. "There is nothing in the duty manager log, which is used to report unusual incidents," she stated. This announcement seemed lacking since so many United employees and pilots reported the UFO sighting to their United supervisors.

The pilots of the United passenger jet being taxied back from Gate C17 were told by United employees of the sighting. One of the pilots reportedly opened a windscreen in the cockpit so that he could get a picture of the flying saucer to show others the proof. The pilot estimated that the strange craft hovered at about 1,500 feet above the ground. That was 400 feet below the cloud deck at the time.

Some witnesses reported that the UFO unexpectedly bolted upwards through the solid overcast clouds, which the FAA reported had a 1,900-foot cloud deck at the time of the UFO sighting. One employee who personally witnessed the flying saucer leaving said that it was like somebody punched a hole through the clouds in the sky. The witnesses said that they had a difficult time seeing it shoot upwards due to its tremendous speed, as if inertia had no influence over the craft. The craft left an empty hole in the clouds that remained clearly visible for close to three minutes. The FAA latter claimed that lights played tricks with witnesses' eyes that night, but the airport ramp lights had not yet been turned on. The only light capable of making a whole through a cloud would be a laser, which would evaporate the droplets and cause a window. O'Hare has no need for burning lasers with twenty foot in diameter beams, so that reasoning by the FAA was not thought out so well. The hole in the cloud was said to exist for up to fourteen minutes by some eyewitnesses in the parking lot area at O'Hare. This suggests that the object that made the hole was super heated or otherwise radiated (possibly microwaves) energy that would have to be on the order of 9.4 kj/m³.

The hole in the cloud, caused by UFO, has been seen many times throughout history. For example in a declassified Army Air Force Intelligence report stated in July 10, 1947 at Harmon Field, Newfoundland, Chief Mechanic for Pan American Airways (John E. Woodruff) saw a "translucent disk like a wheel traveling at a terrific speed and opened the clouds as it went through the air." Then he stated, "the object passed through and cut the cloud leaving a gap where you could see the blue sky."

Also a formerly classified U.S. Government document report (Oct. 28, 1947) was released under the Freedom of Information Act states some commonly listed features of flying discs: "The ability to group together very quickly in a tight formation when more than one aircraft are together; evasive action ability indicated the possibility of being manually operated, or possibly by electronics or remote control, and under certain conditions the craft seems to have the ability to CUT A CLEAR PATH THROUGH CLOUDS."

On March 9, 1977 a club master and professional golfer in Ayrshire, Scotland was the eyewitness to a similar cloud-cutting incident. He and his friend report seeing an odd light hovering over the seventeenth tee, no higher than a telegraph pole. Its glare was so powerful, that nothing behind it could be seen. After hovering for four minutes, the light suddenly bolted upwards into the low clouds, leaving a clear hole in it as it passed through. This opening in the clouds remained for several minutes despite the winds moving the clouds noticeably.

On December 6, 2002 in Ventura, California, two college professors witnessed a dark object move across the sky, then it stopped and hovered. It changed shape from disc to an oval. It soon shot through the clouds cutting a visible hole in which the blue sky above could be viewed.

Elizabeth Isham Cory, a person doing an internal FAA review of the air-traffic at the time of the O'Hare UFO sighting said the weather might have been a factor. "Our theory on this matter is that it was a weather phenomenon," she said. "That night was a perfect atmospheric condition in terms of low cloud ceiling and a lot of airport lights. When the lights shine up into the clouds, sometimes you can see funny things. That's our take on it." This reasoning is interesting to say the least. She says that it was night, yet it was 4:30 p.m. and the sun had not yet set. The airport lights had not yet been turned on. This kind of explanation only contributes to feelings of disbelief and cynicism of the witnesses. It seems so unreasonable as to be ludicrous and begs the question, how could someone who did not even see this UFO come to such a conclusion? It is obvious that if commercial pilots and commercial aircraft mechanics said that they witnessed a UFO, then it can be concluded that it's true. Sigmund Freud once stated this about reading too much into things, "Sometimes a cigar is just a cigar." Meaning that the FAA was reaching for many possible complex explanations when there was only one, they saw what they saw.

Chapter Two: Brazil UFO Crash, November 2006

On November 25, 2006 a Brazilian UFO Magazine editor Gevaerd got an interesting and unusual call from Bahia. The caller was Francisco Baqueiro, who is a psychologist and magazine consultant on alien abductions. He said that he had photographed with his cell phone an object that he believed to be a perfectly disk shaped flying object that had been loaded upon and transported with a semi truck. The saucer itself sat upon a tractor-trailer. The truck had no identifying marks as they traveled a road in the Brazilian state of Bahia.

3.   Photo of Brazil UFO on trailer. (At http://scwbook. blogspot.com/)

4.   Photo of Brazil UFO by unknown photographer. (At http://scwbook.blogspot.com/)

5.   Photo of Brazil UFO by Roberto. (At http://scwbook. blogspot.com/)

He took the picture on Tuesday November 21, at 7:16 am. Baqueiro stated that he heard of the crashed UFO being transported one day earlier when an informer from an intelligence service called telling him about an unknown artifact that crashed and was recovered in the location of Bahia, south of the capital Salvador. The intelligence informer told him that the UFO was being transported from the crash site to a place in City of Feira de Santana, a city that was 110km from Salvador. They took it on the BR-324, a two-lane road with constant traffic.

Even though Baqueiro was very sick from his battle with kidney disease, he decided that this was too intriguing and that he had to check this out for himself, for this, we are indebted to him. On Tuesday, just after 6:30 am, the therapist and his wife spotted the semi truck carrying the disk-shaped artifact and they began to follow it in their car.

Baqueiro saw that there were four federal police cars escorting the truck. Two police cars ahead of the truck and behind it. He also said that the truck moved very slowly down the road, and that the saucer on its trailer was so large that the truck had to drive down the middle of the two lanes, slowing traffic all around it. Oftentimes the edges of the saucer would scrape the trees on one side of the road. He noted, "that was certainly too heavy, making the truck go so slow." The transportation of an uncovered and exposed UFO seemed baffling to him, but perhaps others had control of it before formal investigators could take charge of it. "Even with so many cars coming from the opposite way and the other ones behind the truck, very few people seemed to care about the scene."

They saw the truck stop at Posto Phoenix, a gas station along the BR-324 road, just 60km from Salvador. Baqueiro stated, "I tried to enter the gas station in order to better check what was going on, but was prevented by the federal agents."

After the Federal officers stopped him from entering the gas station, he drove across the street and parked along the road to get a better view. He pretended to be talking on the phone and he took a single shot of the object on the trailer. He was afraid to arouse the officers' suspicions if it was too obvious he was taking photos of the artifact. He stated, "I didn't want to try others because I didn't want to catch their attention." That was when Baqueiro sent the photo to a Brazilian UFO magazine using his cell phone. The people at the magazine uploaded the photo onto a computer and enlarged it to get a better idea of what they were looking at. Editor Gevaerd and others analyzed the photo. They were all surprised by what they saw, but a few of the co-editors were a bit irritated that it looked similar to an earlier photo that surfaced on the Internet where a saucer shaped object was being carried on a trailer in Brazil, but this one was different. This photo that Baqueiro took was extremely clear revealing details of the surroundings as well. As for the Internet photo that is similar, no one came out and claimed responsibility, nor was there any information with it except a date and a location being Brazil. Francisco Baqueiro has been a consultant for the Brazilian UFO magazine for many years. Baqueiro's specialty is in alien abductions. He latter went back to the gas station two more times, but the employees there were clearly too scared to talk about the UFO on the trailer. Now Baqueiro has focused his attention on finding the alleged crash site of the UFO and find out where the object was taken. He stated that, "all I know is that the crash was in a sugar cane plantation, but the object was nearly intact." This region of Brazil is littered with oil refineries, but he said that it's impossible that this was a lid for water or fuel tanks. He had years of work with Petrobras, the largest oil company in Brazil, so he was confident that he would recognize such a part right away, and that they never were the size of the saucer on the trailer. He also denies that the object could be part of an attraction at a theme park for children, since he has also worked at the largest park in Bahia.

Strangely enough, when the news about the UFO crash was seen on the Internet, several UFO researchers suggested that it could be a cover for a large container that they were transporting in separate pieces. That would explain why it was being transported in broad daylight without being covered up. Yet, had they covered the object, it may have gotten even more attention than it did. Perhaps they just didn't have a cover at the time, so refusing to miss their deadline, they began transporting it.

The story takes an unusual turn when the Brazilian UFO magazine got an email from a nuclear engineer named Luiz Carlos Pires. He sent a message that he was positive that the disk-shaped object was a cover for a factionary distillation machine at a nuclear plant. Let me explain why this last assumption of the UFO being a cover for a factionary distillation machine is not possible. FIRST, any employee at a nuclear plant is sworn to secrecy and would not call or send emails revealing sensitive intelligence of nuclear facility parts that may jeopardize any nuclear program that the plant is undertaking. SECOND, factionary distillation is used for nuclear reprocessing which separates any usable elements like uranium and plutonium from fission products and other matter in spent nuclear reactor fuels. The overall goal is to recycle the reprocessed uranium or place the elements in new mixed oxide fuel (MOX). This processing can be used to obtain plutonium for weapons of mass destruction. Now does this sound like classified material yet? So why would a nuclear engineer risk compromising the security of the nuclear facility? Not to mention, if it is a lid for a factionary distillation, then why does Brazil need weapons of mass destruction? It is clear that Brazil has pursued a covert nuclear weapons program in response to Argentina's program, yet to email this information out to a large Brazilian UFO magazine which will obviously publish it, would be considered as an act of treason against the Brazilin Government for revealing Top Secret Classified information. Therefore this explanation by the nuclear engineer is clearly impossible. No nuclear engineer would risk such a thing. The email was never responded to nor investigated further, but merely taken as the truth.

It seems more possible that the person who emailed the UFO magazine was just a government employee who was told to debunk or try to cover up the story ASAP. This explanation seems more likely and is a typical governmental response to such news media leaks.

As for the other photograph of the saucer shaped object on a trailer truck, it is clearly a similar instance with what appears to be the exact same trailer. The saucer itself is different in that it in only a single layer, where the sighting on November 26, 2006 was of a saucer-shaped craft with one small layer on top of a much larger layer.

Another strange sighting occurred in Alagamar, Brazil in November 26, 2006. The reason for mentioning it is that the clear photographs that were taken were obviously the same type of saucer-shaped craft seen on the blurry tractor-trailer photo, the same Internet photo that had no information about it. The photographs without a doubt show a single layer saucer in a lightly cloudy blue sky. The person that took these clear photographs is Roberto Di Sena, former military. He said that he took the pictures between 8:30 and 10:00 am on November 26, 2006. It was at a location called Alagamar, state of Reo Grande no Norte, Brazil. Di Sena didn't see anything strange in the sky and took several pictures at random, deleting many later. Luckily he didn't delete these clear beautiful photos of the UFO. He used a Pentax Optio 60 6.0 mega pixels. The object from his photos clearly matches the Internet photo that had no information accompanying it. This backs up the story that the blurred photo of the UFO on the trailer is real.

Chapter three: Shuttle Atlantis UFO Encounter, 2006

Imagine if you will, the Shuttle Atlantis orbiting Earth and being followed closely by something NASA called, "a UFO." Well that's what CNN, Larry King Live, The Seattle Times and ABC news reported. Atlantis was launched into space on September 9, 2006. An intense twelve-day mission and a stressful twenty-four hours with what NASA said were three gray metallic circular-shaped UFO's changing positions around them in orbit. These objects flew around the shuttle, racing from the front to the back, quickly changing positions and then just hovering for a while. Latter NASA tried to say the UFOs were objects that had unknowingly floated out of the cargo bay. It is impossible that objects could have floated out of the shuttle cargo bay and had the ability to increase speed, to pass the shuttle and be visible in front of the nose area, as the recorded NASA conversations show.

6. Photo of Atlantis UFO close up. (At http://scwbook. blogspot.com/)

7. Photo of Atlantis UFO farther. (At http://scwbook. blogspot.com/)

Fox News showed clear video of not one, but four metallic ball-like UFO's near the shuttle. These objects appear perfectly round and occasionally move from one location to another around the shuttle. If it was debris, it should not be able to move so suddenly and still stay in perfect orbit with the shuttle. The objects seemed to zoom to other position around the shuttle quickly and then hold their position for a while as if they were curious about the shuttle. NASA admitted they often see objects like these through the years and even chased after them a bit, but none have ever gotten the recognition that this mission has gotten, yet they said they have never been able to find out where these circular-shaped objects came from. NASA eventually came upon the conclusion that the tiny ball like UFOs didn't pose any threat to the shuttle.

NASA delayed their planned landing of the space shuttle Atlantis by more than a day because engineers spotted several unidentified flying objects matching the Atlantis's orbit perfectly without deviation. On CNN, NASA declared it as a UFO and then latter claimed it was an object that may have accidentally flown out of its cargo bay. NASA kept repeating that it was "a UFO" rather that what Fox News video reveals showing visibly at least 4 or more dull grey ball UFOs. One object was estimated by FOX news to be close to the size of a small house. An object of this size could not fit into the cargo bay of the shuttle, but when NASA engineers' scrambled for an explanation, the best they could think of was that it came from the cargo bay.

NASA engineers became frantically worried that the item that floated out of the cargo bay may have been a crucial element to the shuttle and may cause a severe accident on reentry as in 1993 when the space shuttle Columbia disintegrated on reentry, killing all seven astronauts aboard. Many UFOlogist claim that the Colombia collided with a UFO on reentry, but that has yet to be proven.

NASA engineers believed that the object might have shook loose from the shuttle during the firing of jets in preparation for landing. The shuttle crew was ordered to do an additional inspection, something the Atlantis crew already had done twice earlier. An inspection is tedious and time consuming, because the crew and those who break down the data that is sent back to Mission Control are looking for the tiniest of cracks that could allow damaging heat to penetrate the shuttle during re-entry. This inspection the third time seems to have been done to keep the crew of the shuttle busy on their work and keep their minds off the UFOs flying outside their windows.

NASA even decided that the UFO may be a serious threat to Atlantis and they circled the date November 11th to be the day that they would be forced to launch STS 300, shuttle Discovery in case the Atlantis crew needed rescuing in orbit. This plan may have been made because the UFOs themselves were dangerously close below the Atlantis shuttle, preventing it from leaving orbit or attempting a landing, heightening the risk of a collision. The Discovery would have raced to be launched with a skeleton crew. It would have rendezvoused with Atlantis, orbiting as close as 50 feet away, each shuttle's open payload bays facing one another at a 90-degree angle. A rope tied to an astronaut would carry two spacesuits to the Atlantis, allowing them to escort them two at a time back to Discovery. Then once empty, the Atlantis would have been sent plummeting into the ocean by remote control. Fortunately this didn't happen.

Wayne Hale, the NASA space shuttle program manager stated, "The question is what is it? Is it something benign? Or is it something more critical we should pay attention to?" Then Mr. Hale continued. "We want to make sure we're safe before committing to that critical journey through the atmosphere."

Astronauts aboard the shuttle Atlantis took numerous photos and video of the UFOs and described it (still referring to the many UFOs as one) to Mission Control. "It's fairly small…it was departing away from us, maybe one or two feet per second," Jett announced over the radio. "It wasn't rising or falling…it was definitely moving away fairly quickly."

Mission control saw the UFO the size of which would not be disclosed by NASA, with a video camera in the shuttle's cargo bay. The object, which circled the Earth in the same orbit as the shuttle several times, was in a circular-shaped metallic ball, yet CNN video shows the object to be a rectangle shaped brownish object that seems to rotate end over end as it moves across the sky with the clouds below it. This discrepancy of the actual object seems baffling, yet perhaps the object has the ability to change its shape as some UFO's have been said to do, but the most likely solution is that the small metallic balls came from the long brown object and were probes of some sort, scanning, taking readings and such.

NASA Mission Control communicated through a digital downlink to the shuttle Atlantis. They spoke about the UFO's they saw from the payload by camera. In this conversation, they used code words like *rings* and *reflections* and *foil* so that they don't cause anyone to panic who happened to overhear their conversation, like ham radio operators:

> Mission Control: "Do you see the digital downlink?" (Video of four metallic orbs moving along side the shuttle).
>
> Atlantis Astronaut: "Uh…We do see that."
>
> Mission Control: "Okay. That's uh, looking out in front of the orbiter."
>
> Atlantis Astronaut: "Okay, we are seeing three or four objects. Can you confirm that it's just the one moving? That the other ones are just reflections?"
>
> Mission Control: "No, there are, there are three objects. The one you see, you see two rings right there. They are the ones that we got the late tally-ho on. Uh, the one down at the bottom, that's the one we initially saw that appeared to be foil. So we got two rings and a piece of foil. The piece of foil seem to be further in front with the grounding strap on it. It seems to be further in

front of the other two, and the other two seem about the same equal distance from both us and the forward one."

Atlantis Astronaut: "Copy that."

In the next conversation between an Atlantis astronaut and Mission Control, you will see that the astronaut is concerned over the distance of one of the UFO's and if he had the distance of the object, then he and everyone else on Earth could easily calculate the exact size of UFO's using the video from the shuttle. If the exact size were larger than that which would fit into the cargo payload, then it would be definite proof that they were UFOs, but mission control handles this curious astronaut well. For instance, if the objects were a quarter a mile away, that would make the objects much larger than the shuttle itself.

> Atlantis Astronaut: "Yeah, I'm just wondering. How close we are to the trajectory foam by the Soyuz." (He was referring to the long brown-red UFO he saw out his window as the trajectory foam.)
>
> Mission Control: "Yeah, we're asking." (Long pause). "We're going to check on that, but, uh, while we are waiting for that, uh, we were thinking based on our daylight and KU availability, if we can get back to the OBS inspection and maybe use the still cameras to photograph these objects."
>
> Atlantis Astronaut: "Yeah, we understand. We're getting back to OBS." (The astronaut sounded embarrassed).
>
> Mission Control: "Thanks!" (Said in a higher tone, as if he was relieved from the stress the astronaut gave him).

The mission management team had ruled out that something hit the shuttle as it was in orbit. The NASA engineers believed the sensors that indicated "hits" were reacting to the forward control system maneuver, which is part of the preparation for landing. If the sensor alarm for "hits" against the shuttle did sound, and this UFO was in synchronous orbit with the shuttle, it can be concluded that the UFO came in contact with and actually touched against the shuttle Atlantis at least one or more times. If these small gray metal balls were probes, then they might have placed objects on the shuttle to keep track of its location, or getting closer scans of its inhabitants. It may have been these "hits" of the UFOs bumping the Shuttle Atlantis that may have caused NASA to panic into thinking it may have to send up the other shuttle to rescue them.

"It's something that we didn't expect, but it's something that we're taking a real close look at," stated one NASA engineer. NASA then ordered the Atlantis to keep its cameras running all night instead of stowing them ahead of the planned landing attempt as they usually do. This shows their fear of the UFOs and how worried NASA is for the astronauts on the shuttle. For NASA to not be prepared for a contingency can mean only one thing, alien interaction, because NASA has a contingency plan for every possible scenario, but alien interaction is the one that they would be most surprised and confused by.

The Shuttle Atlantis even had a rare conference call with two other spacecraft also currently in orbit. This conversation seems to make light humor of the UFO's that were following the Shuttle Atlantis.

"It's a little crowded in the sky this morning," said Jeff Williams who is a crew member on the international space station that the shuttle undocked from on Sunday after the Atlantis delivered a solar panel addition.

"We were wondering if we had to hire some more air traffic controllers for the increased traffic up here," answered Michael

Lopez-Alegria from the Russian Soyuz space capsule that was launched from Kazakhstan earlier on Monday.

This 2006 encounter with the shuttle Atlantis was not the first by a shuttle. NASA STS-37 on April 1991 had an encounter with a large black triangle-shaped object. It moved to the right, and to the left, up and down. This triangle object moved back and fourth with the Earth as its background (meaning the black triangle was easy to see with the Earth behind it.) This was seen over four times on NASA video and the conversation reveals that they may have gotten contact from it on channels one, two and three:

Mission Control: "Atlantis Houston, about a minute from LOS. Jay, I was about to tell you to let us change the fuse, but let us talk about this one more time. I'll get back to you when we are AOS at 20 36.

> STS-37: "This is Atlantis."
>
> Mission Control: "Atlantis Houston. How do you read?"
>
> STS-37: "Yeah, before you go over the hill, uh, the roll D auto on the left side only made contact on channels two and three the first time. The second time, uh, our third push on it, we got all three channels.
>
> Mission Control: "Copy."

On September 12, 1991 a mass encounter occurred by STS-48 shuttle. This encounter was publicly seen on "Hard Copy" and "Larry King Live." STS-48 had not one, but six encounters of UFO's that appeared to shuttle crew and got recorded on NASA Live coverage.

EVENT 1: Happened between 3:49-4:10 GMT. The camera focus was looking back through the cargo bay at the tail section with Earth filling the top of the screen. Suddenly a series of five orb lights travel radically from behind the tail section comprising an arc shape. Then an extremely bright light appears from behind the engine of the shuttle.

EVENT 2: Happens 9-15-1991 between 20:30-20:45 GMT. In this encounter, the Earth is filling most of the frame with empty space on its right side. Several-lighted orb-like objects are clearly seen traveling slowly across the sky below the shuttle. These objects make sudden changes in direction, at incredible speeds.

EVENT 3: Occurs 9-16-91, at 8:40-9:10 GMT. Here the video shows the Earth filling the screen. Several orb like objects are seen, with sudden spurts of speed.

EVENT 4: On 9-15-91 between 19:00-19:10 GMT. A large bell-shaped object appears, and other similar shaped objects move from the bottom to the top of the video screen.

EVENT 5: (Date and Time unconfirmed). A large tubular shaped UFO moves from the right to the left of the screen. Also a pulsating UFO was at the upper part of the video. NASA later deleted this 14 minutes of video from their archive because of it being too impacting of evidence.

EVENT 6: (Date and Time unconfirmed). A bright object enters the video screen at the left center and shoots across the screen at a high speed. When the video is put into slow mode, the object actually appears to be a white light traveling in a smooth trajectory, but when looked at frame-by-frame the object changes

color from red to green to white and instead of
a smooth trajectory the object travels in short
discrete steps turning invisible in between the
steps.

In November of 1994, Shuttle mission STS-51 encountered
a round metallic orb, similar to the ones the shuttle Atlantis
encountered in 2006. This object was recorded on both NASA
video and photographs.

Another NASA Shuttle that encountered a UFO was STS-
75 on February 25, 1996. This shuttle was filming outside at
a tether line that extended from the shuttle that NASA said
was "twelve miles long." As they filmed, a pulsating doughnut
shaped light appeared. Its entire surface pulsated with white
light covering its entire surface and then turning dark and
then light again. It seemed to throb much in the same way a
heart grows as it beats, then gets smaller. It was recorded flying
past the end of the tether line slowly. Its size was estimated
by UFOlogist to be two to three miles in diameter, measured
against the twelve-mile long rope said to be extending in space
and was not attached to the shuttle. The video can be seen at
http://www.youtube.com/watch?v=fwSmb1LCruI and is very
remarkable. At three minutes into the above video, you will see
not one, but over fifty pulsing, glowing disk-like objects flying
past the strand of broken tether. They say the tether was 900
miles away from the shuttle when this video was shot.

The STS-75's mission was to deploy an experimental tether
into orbit. The experiment was labeled the TTSS-1R (Tethered
Satellite System) and it was suppose to make electricity by
gathering it from the Earth's magnetic field. Sadly, as they began
to unravel the tether, they encountered a problem and the line
broke, but five hours of data were recorded from the experiment.
It is very likely that one of the UFOs seen in the video above
may have accidently crashed into the line, not expecting it to
be there. In the video, more and more UFOs appear as if they

were just informed of the tether and were curious about it and came to check it out for themselves. What latter happened to this twelve mile long tether in orbit is anyone's guess.

According to my sources, NASA no longer gives out a live downlink from shuttle missions, but instead uses a delay similar to that used by talk radio stations to censor offensive remarks. So NASA is cleansing the download links from the shuttle missions, and deleting sections of the official archives that are not in the interest of the taxpayers who fund NASA. It is obvious that further inquiry into this matter is necessary, which will of course, be ignored by the US government. It is true that lighted objects could be explained as debris or ice crystals lit up by the sun, but I cannot imagine how that could explain these objects traveling in opposite directions. Clearly thruster plumes would propel debris in one direction only. It is obvious that these orbs, bells and tubular objects are intelligently controlled UFO's that have been and will continue to study the shuttle on every launch as well as the ISS (International Space Station). We like to study the things around us, just as aliens too enjoy learning of new things, something that governments have difficulty coming to grips with, even with floods of evidence.

Chapter Four: Argentina Police officer abducted, 2006

Imagine you're a police officer patrolling the streets on your police issue motorcycle through the beautiful mountain roads of Argentina, when you pull over and are confronted by two aliens and one other tall creature, that take over your freewill and begin issuing you orders, just so they can better learn about humans by performing tests on none other than…lucky you. Well, such an occurrence happened to Pucheta one evening and other officers from his station have also confirmed seeing the UFO lights around the mountain area.

8. Photo of police officer found at roadside. (At http://scwbook.blogspot.com/)

9. Photo of weapon parts and cell phone on ground. (At http://scwbook.blogspot.com/)

It took place in General Pico, La Pampa, Argentina. It was March 2, 2006 when Corporal Sergio Pucheta, a local police officer unexpectedly came up missing from his route through the high mountain pass. The officer had been missing for a total of twenty hours before he was finally found. His last statement he made on his cell phone was him reporting a "strange situation," but his connection was cut off. It was only a matter of ten minutes before there were backup police officers scouring the area for him. They found his last location where he apparently called from, yet they were baffled by what they found. The police officer's motorcycle laid on the ground sideways, with his disassembled weapon sprawled across the ground along with his helmet, radio and cell phone.

The next day the police led an intensive search over the entire mountain where the officer had vanished. Finally they found him at 4:30 pm sitting down on the side of a dirt road, crouched in a fetal position. They found the officer between the small towns of Trili and Quemu Quemu, approximately 20km from the location where his personal items and motorcycle were found a day earlier. One of the local farmers happened along the road when he noticed the man sitting down with his arms wrapped around his bended legs and his head hidden between his chest and legs. The farmer said that when he found the officer, he appeared to be in a state of nervous shock.

As soon as other officers came to help him, Corporal Sergio Pucheta began telling them about his sighting of the two small humanoids with glowing red eyes, which apparently communicated by using telepathy, and gave him orders. He said that he tried to escape from the two aliens, but could not remember much after that. He was even confused about how he had ended up 20km away on a desolate and rarely traveled road. He said the two aliens followed him around and gave him orders telepathically.

Earlier records of Corporal Sergio Pucheta patrolling the area regularly had turned up reports of him seeing strange lights and he had tried to film those lights on many occasions so that he could show his fellow officers proof of his reports.

For some odd reason, the officer that they saw fit to carry a gun and protect and serve the people was placed into psychiatric care. The doctors say he was suffering from headaches and his hands itch, which doesn't justify the psychiatric care since hand lotion and some aspirin can take care of those symptoms.

Corporal Sergio Pucheta said that he had been driving his Honda 125cc motorcycle when he finally got to the summit area of wilderness called "Las Canas." At this point he saw a red light that appeared to be the taillight of a car, but when he reached the location of the source of light, he found nothing unusual and the light had vanished. He felt confused and pulled his motorcycle to the side of the road and got off to look around and explore the surrounding area. When he found nothing unusual, he turned back and began walking back to his motorcycle. As he stood next to his bike, the red light returned and floated in front of him. He was awestruck by the luminosity and dismayed at why he suddenly could not move his hands. Then he watched as the light rose high into the air, but something suddenly caused him as considerable headache and eye pain. He panicked and tried to run away from the light to ease the pain and turned and bolted toward a nearby field of corn. The officer never remembered dismantling his police pistol or the walkie-talkie, both of which were found sprawled out on the blacktop road a few feet from his motorcycle, which sat on its side. The officer said he became fixated on the light, freezing like a frightened deer in the headlights of a car. He said at this point that he could not feel his arms and hands. We can assume that at this point he was under control of these alien-like creatures, dismantling the gun into many pieces and the walkie-talkie, taking off its antenna and pulling out its battery. If he did that himself, then this was the moment in the story in which he had first contact with the aliens.

This brings us to his cell phone found on the ground. The cell phone was found still in working condition, but with a few twists to it. The cell phone's memory had been wiped clean of all the stored information and phone numbers that Corporal Sergio Pucheta had put onto it, except one. The last call the officer made, he apparently didn't use the phones memory to dial, but instead pushed in the numbers one by one himself. The means the memory was deleted before he tried to dial for assistance. Not only that, but he said the walkie-talkie didn't work, indicating again that a possible massive energy source was close enough to both wipe out the cell phones memory and disrupt radio communication. If instead of making a cell phone call, he had begun taking video or photos of the aliens, then he would have some evidence to prove his story, assuming the power surge came before that call, but then hindsight is always twenty-twenty.

Corporal Sergio Pucheta was pursued into the field by the two aliens that appeared to not even touch the ground, but instead levitate over it gracefully. He said the creatures were somewhat shorter and thinner than him in stature, and were also semi-transparent. They had overly large heads and their eyes were clear and red. He said that the two aliens were obviously using his mind for performing scientific tests. When he had reached the cornfield, he saw a third creature, like nothing he had ever seen before. It was much taller than him and was chewing the corncobs and breaking them in half. He felt too scared to walk past the creature and only stood there gaping at it. It began to approach him step by step. That was when he heard the two aliens tell him, "either go forward or backward." He then walked right past the tall creature. The fear he had inside him suddenly left him. Another officer asked him if the third figure was some farm animal, Pucheta replied, no because it was much taller than me and much larger in size.

Two weeks earlier, another officer, a friend to Corporal Sergio Pucheta, was driving along an access road to Agustoni and he reported seeing a strange light in the middle of the road. Believing it was a car he flashed his headlights at it, but the light didn't move out of the center of the road. Finally when he was about to overtake the light, it abruptly move to the left of the road, then to the right and then straight up and out of sight. This left the officer feeling confused at what happened, but he filed a report on it at the police station.

Medics drove Corporal Sergio Pucheta to the Centeno Regional Hospital where Dr. Covella assisted him. The officer was looked over by many specialists and was tested for many things, finding finally that his physical state was perfectly normal and hydrated. Dr. Covella said that the officer merely appeared to be exhausted from lack of sleep, which Corporal Sergio Pucheta confirmed by continually asking to being let alone so that he could rest.

As rare as it is, of having a police officer personally witnessing aliens and their spacecraft, it is even more unusual that a year latter it happened again. A year latter and 411km away from the city of Trili, came another alien sighting by not one, but four police officers. In this sighting they too experienced paralyzed parts of their body and said the aliens were short in stature.

The UFO sighting happened on November 10, 2007 near the city of Irene, Coronel Dorrego district, which caused the local residence to question the rationality of their police force. It all began on a Wednesday at 1:30 am when two police officers coming out of the Oriente station began to investigate the rural area of Irene, but were directed to the property of Felipe Fernandez. The police officers were on a typical patrol of the neighborhood when officer Orellano suddenly gets out of the police truck to take a closer look at something.

Officer Bracamonte stayed in the truck trying to get a new phone card into his cell phone. He saw a small light like that of another vehicle coming toward him. The light came closer and brighter. It didn't seem to bother him at all, but then when it was at a distance of about 10 meters away, he saw a shape moving and his first thoughts were that it was a dog wandering through the headlights of the other vehicle. He clearly saw the silhouette, and realized that it was a person not a dog. The person was standing about 80 cm high. He saw the person had a large head, large round gray eyes and had a greenish tint to their skin. He said, "I tried to dial my fellow officer's cell phone, but when I pressed the 1 and the 5, my hand remained sort of static. I could see that three creatures came out of that vehicle or craft, two similar to the first and a fourth with a slightly more robust appearance."

At the same time his partner was also seeing the same exact thing and he yelled out, "What's going on?" The four aliens panicked and bolted for the ship. They entered the craft and it quickly rose up and shot off towards the north. In its wake the craft left a white light trail and behind that was a green halo of light. The officers smelled a powerful odor that is most similar to that of sulfur or gunpowder. Also the ship made a sound similar to thunder as it raced away.

(Note: the smell of sulfur may be linked to the atmosphere the aliens' breath inside their ship. I don't think it is linked to the ships propulsion system. Phil Schnieder's statements lead me to this conclusion. Refer to Underground Base chapter.)

Bracamonte had no idea what was going on, but when his partner sat back inside the truck, his hand still had no feeling whatsoever in it. Also for two more hours he had trouble with his eyes, because tears would not stop flowing.

When the two officers got back to the police station, they reported to the others what they had seen. They soon found out they were not the only ones that morning to witness the event.

Two other police officers, sub-lieutenant Santiago Walter of the Aparicio police station said that when they were on a routine patrol just two km away from the city of Irene, they saw something unusual. It was between one and two in the morning. They were driving out of town (Nicolas Descalzi) and headed in the direction of Aparicio when they saw a large white light that suddenly began to cover their truck.

He said, "at that time the truck stopped dead and I tried to place a call on my cell phone, but there was no signal. My partner, Carabajal, asked me to look up. That's when we saw a powerful light that appeared to stop above our pickup truck and then vanished in a flash." Neither officer could explain what they had witnessed.

Local residence of Oriente confirmed the officer's reports saying that they had seen a strange glow flying over the towns of Irene and Aparicio between one and two in the morning. A married couple had been driving from Marisol from Oriente at the same time and said that their vehicle was covered in light that seemed to quickly travel from Irene before hovering over them.

All these sightings bear many similarities to the Travis Walton abduction case in America made famous in a book and film called, Fire In The Sky. It took place in November of 1975 when a group of six people employed as tree-trimmers began driving home from work in an old pick-up. They drove through the Sitgreave-Apache National Forest in Arizona, USA. The group saw the lights of a flying saucer hovering over the trees and stopped the truck to get out and get a better look at it. It was hovering about fifteen feet above the pine trees. Travis Walton wanted to get a better look so he walked directly below the craft that was now hovering in the grassy field that they stood. His friends protested and tried to get him to come back, but he didn't listen to them. Instantly Travis was knocked to the ground by a wide blue and white beam of light. When the men in the truck witnessed this, they panicked and drove off down the dirt road believing that the UFO had killed Travis. Latter when they got their wits back, they drove back out to the location where they had left Travis, but they could not find any sign of him. Five days latter with the newspapers blaming the group on killing their friend Travis, he was found on a rarely traveled country road, with no memory of what had happened to him over those few days. Latter the group all underwent police lie-detector tests, all of them passed except one, which was undetermined.

In conclusion, the police officers will have to endure loads of criticism as law enforcement officers, that is, if they are lucky enough to be allowed to keep their jobs. When Corporal Sergio Pucheta (March 2006) described the two aliens being followed by a much taller creature, it is clear that this taller creature is another intelligent alien species, meaning, yes, there are more than one species of aliens out there. Read the chapter on Area S4, to find out exactly how many exist. Also note that this sighting is on top of a mountain, where there have already been many sightings of UFO lights over the years. Many UFO sightings every year around the world, take place near tall mountains or volcanoes. As they approach the mountain, most disappear suddenly. This can only be explained by them flying through some digital shield into some underground base. (For more info in these underground bases, read my chapter on it at the end of my novel, Dragons of Asgard.) Understand these are alien driven craft and not ghosts or gods that can travel through mountains. They are flesh and blood like you and I. If the police officers involved had been lucky enough to get some video or photos of the aliens on his cell phone, things might have turned out differently for them. But sadly all Corporal Sergio Pucheta has is his recollection of what happened on the mountain and the witness testimony of several other officers who have seen the strange UFO lights in the area. The police officer in the March 2, 2006 sighting clearly was overly sensitive and prone to being afraid of things. It sounds like his fear controlled him more than the aliens did.

Chapter Five: Apollo 20 Alien Technology Retrieval Mission-2007

On April of 2007, several videos apparently taken by NASA have emerged showing a city on the moon and a close up of an alien cigar shaped vessel in Deporte crater. William Rutledge, who claims to be a former US astronaut that took part in this mission, released the videos. These videos were from the Apollo 20 mission. Apollo 20 was a covert Apollo missions to the moon to retrieve ancient alien technology. I'll be the first to admit this really seems ridiculous on first glance, but if you do a little research into William Rutledge's story, then you begin to see that NASA has been hiding the truth for a long time, editing what we are allowed to know and not know. When researching this story, I hoped to learn the truth about the covert missions to the moon. What I learned instead stunned my imagination, strengthened by beliefs and amplified my fears.

10. Photo of city on moon. (At http://scwbook.blogspot.com/)

11. Photo of right side of city on moon. (At http://scwbook.blogspot.com/)

12. Photo of ship on moon at Deporte Crater. (At http://scwbook.blogspot.com/)

13. Photo of close up of ship on moon. (At http://scwbook.blogspot.com/)

14. Photo of face of EBE found on ship. (At http://scwbook.blogspot.com/)

15. Photo of JFK memo of November 12, 1963. (At http://scwbook.blogspot.com/)

Rutledge does show that he has an extreme knowledge of Geology, Chemistry, and space exploration, which can be seen in his vocabulary that he uses during his interview for a UFO magazine. Another reason that Rutledge seems credible is that he didn't gain money from the UFO interview, nor has he so far. So what could be his motive? Perhaps it is to bring out the truth about the existence of alien technology on the moon.

William Rutledge is an eighty-year-old man currently living in Rwanda. He was born in 1930 (Note: Rutledge is the same age as astronaut Edgar Mitchell is a brilliant scientist and an all-American hero who was born in September 17, 1930.). He claims to be an astronaut that traveled to the moon on a mission called Apollo 20. Now Apollo 17 was supposedly the last mission for NASA. At that point, NASA cancelled the next three missions, claiming that their budget currently was not capable of funding such an exploit. William Rutledge said that NASA's Apollo 14 mission took photographs as they flew over the polar region of the dark side of the moon, recording numerous ships, odd shaped buildings, and ancient towering cities, most appearing abandoned for million years. As I investigated, I found that all Apollo missions from 14 to 17 flew over that polar region of the dark side of the moon. Each of those missions landed within a few hundred miles of each other, indicating that something in the area was of high scientific value. The main mission of Apollo 20 was to explore a mother ship long since abandoned, but discovered in Apollo 14 photographs. The mission took place near Deporte Crater, where a cigar shaped object is clearly visible even with the naked eye, in these NASA photos at:

1. http://www.lpi.usra.edu/resources/apollo/
    frame/?AS15-P-9630

2. http://www.lpi.usra.edu/resources/apollo/
    frame/?AS15-P-9625

Before investigating William Rutledge's claims I first looked at the NASA panoramic photos above. It took me about three minutes to spot the cigar shaped ship poking oddly out of a crater area in the right third of the photo that he speaks about. It is elongated and appears to have a cockpit area with windows close to the nose area. I didn't feel this was enough evidence to prove that he was being truthful. Perhaps he found the same photo and made up the whole story? I dove deeper into the photo. Photograph 9630 got my attention first. I explored the far left of it, peering at its dark shadows and those unusual reflective objects showing through them. That's when I first saw a silver odd-shaped building clearly visible with the naked eye and without changing the photo in any way. To learn more of what I found, refer to the chapter on Moon discoveries. Needless to say, I learned that William Rutledge was telling at least one truth; an abundance of alien technology does exist on the area of the moon that he suggests. Abundance meaning there is so much there that it is impossible to count it all.

Now lets explore his story. Using the panoramic photo that he speaks about, and enlarging it 500% using any photo program, many alien objects will become clear to the naked eye, but lets not stop there. Instead, let's use an amateur CSI technique for digital photos. This technique is quite simple, take the enlarged photo, looking through a digital camera while the photo is on the computer screen. This causes images that were blurred by a photo program to once again become semi-focused, bringing it back up to 80%-90% of the original image. Yes, this is re-digitalization, but once it's in focus, you will be amazed, nay, awestruck at what you see before you.

Rutledge claims that a secret joint space mission, American-Soviet collaboration began in August of 1976 in search of alien technology. Is it possible? After looking at those photos, using the amateur CSI techniques, I found his story to be credible. Numerous structures, many with lighted windows can be found in the photo. When you see a dark area of the moon all covered in shadows and a spherical or geometrical shaped building in the shadows with a row of lit windows across it, you can assume someone's home.

Since April of 2007, a man claiming to be William Rutledge has been constantly disclosing loads of information and video footage along with photographic materials on the Internet site www.youtube.com. His user name on YouTube was *"Retiredafb,"* and his video of the moon city got over 1.5 million views before, someone (smells like NSA) mysteriously hacked into his account deleting all but one, 4 second video. This hacking into his account only adds to the proof that William Rutledge is the real deal. At last I viewed (before it was deleted) one of his videos had over 1,500,000 views to it and was quickly on its way to 2,000,000. He needed that account because the videos had such a high view number that they were then placed on the most viewed home page of YouTube, because of this increased publicity for him, I find it improbable that he himself would delete the account. He is eighty years old so putting them on again is a tedious ordeal and someone else may have to do it for him. It is possible however that he got so much exposure that the government sent professionals to erase his videos and block him access and send out emails saying he admits to lying. Yes, they sent out messages claiming to be him. Since then, he has created a new account called *"ValValientThor."* Its curious that he chose this name, because Val Valiant Thor is supposedly the name of the first alien contact where a human-like alien landed his spaceship in Washington near the Whitehouse to meet with the US president back in March of 1957. To find William Rutledge simply type "Apollo 20" on YouTube or go to his new site at:

http://revver.com/video/624297/apollo-20-alien-spaceship-on-the-moon-csm-flyover/

Let's review four of his videos. His Youtube.com account has only one video left since someone broke into it and deleted all of them, but some of his followers added his videos to their Youtube pages. William states, "To those who did the job; congratulations, you are well paid." He is referring to those that broke into his Youtube.com account. He made a new account at Revver.com.

1. Title: Apollo 20 Alien Spaceship on the Moon CSM Flyover. Found at http://revver.com/video/624297/apollo-20-alien-spaceship-on-the-moon-csm-flyover/ Video length is 2:39. In the beginning of this video, it shows the inside of the CSM module where you plainly see a flag sticker that is half American and half Russian. Below this sticker is the Apollo 20 mission emblem, which has two small space ships carrying away a long cigar shaped ship from the moons surface. Note the moons surface in the badge, looks like it has an unusually shaped white building on it. 43 seconds into the video you will begin to see the moons surface, through what appears to be a circular viewfinder. Then you begin to see that it is not just over the cigar shaped alien ship in Deporte crater, but it is a mere few feet from its top. The camera is scanning the top surface of the ship from end to end, revealing various deeply carved hieroglyphics in its outer hull. At 1:24 you can see impact holes on its surface. I don't believe these

to be caused by meteors, but looks like some laser or particle beam burnt through it. At 1:25 it gets really detailed of the top hieroglyphics. It looks Mayan in appearance. At 1:38 you see some hieroglyphics, then there is an area that has long lines, much like the chin area of a blue whale. At 1:40 into the video you see the best view of the Hieroglyphics yet. At 1:52 you can see a hieroglyph that looks like a person wearing a headdress and glasses. This is an amazing close up video of the ship.

2. Title: Apollo 20 E.B.E. Mona Lisa TV unscheduled transmission. This Video is at http://revver. com/video/797542/apollo-20-ebe-mona-lisa-tv-unscheduled-transmission/ and its duration is 3:13. It starts out inside the module where you can see a person dressed in a NASA white jumpsuit. At 25 seconds in you can see outside the module window at some structures outside, it looks like another module sitting on the moons surface. At 1 minute into the video you will see the EBE, which looks like a woman laying down, her face appears chapped, yet glossy at the same time. Possibly the face is covered in a wax-like coating. I notice that between the eyes above the eyebrows is an unusual bump. This bump is a lot like the Asian Buddha statues that have a similar bump. It may be a possible third eye, but it is never opened in this video. Her neck and upper chest is covered with an unusual gold metallic towel. Her chest and pelvis appear naked, and glossy. (Note: I see the guy playing around with a camera, flipping it. There appears to be no weightlessness or little difference from that of Earths gravity-needs more research on why? Seems the gravity in this location

of the moon may be same as Earths.) Good close up of face at end of video.

3. Title: Apollo 20 E.B.E Mona Lisa 16 mm film. Length is 6:45. It is found at http://revver.com/video/797241/apollo-20-ebe-mona-lisa-16-mm-film/ . This film starts out with a close up shot of the alien girl they call Mona Lisa. It's a close up before they removed the odd wax like covers of her two eyes, as well as it keeping her mouth open and the bump over her eyes was covered in the wax like coating, possibly indicating a third eye? Condition of the face looks like it was frost bitten, as if frozen before death. They have a paper like object with alien writing on it. It has been placed in a zip lock clear plastic bag. It may be possible that if this paper is legit, that a digitally enhanced photograph of it can be taken from this video, giving us the ultimate evidence of an alien species. This would be the best way to prove that Apollo 20 actually took place. This piece of paper with black alien writing scrolled in a cursive like manner could very well be the next Rosetta stone. It is clear on the video, so it should be easy to get a copy made and printed out to analyze. 2:45 into the video you see a close up of the material of a blanket or towel of hers. This covering close up looks like a circuit board with blue, light blue, and red conduits (very cool material and must have other functions besides being mere clothing. 3:02).

4. Title: Apollo 20 Legacy Part 1 The City. Length is 5:02. Found at http://revver.com/video/624300/apollo-20-legacy-part-1-the-city/ . This is my favorite of all his videos. The city in the background looks so cool, as well as the giant building on the far left of the video that they call the Cathedral. Apparently

the Cathedral was the only intact part of the city. The video is being taken from the rover. At 1:30 into it, you see the Cathedral building and it looks like something from a sci-fi movie. The detail is incredible. They zoom in on the building to see it close up. Then they scan over the many buildings of the city, from end to end.

A magazine interview conducted by Luca Scantamburlo was done through a YouTube email account in May 25, 2007. He says his name is William Rutledge, born in 1930 in Belgium. He is an American citizen, but is not accustomed to speaking in English anymore since 1990 when he moved from Europe to Rwanda. He has learned Kinyarwanda and uses French and sometimes German because Rwanda is a former Belgium-German Colony. He said that he moved to Rwanda because of a woman he met. He says, "NASA didn't employ me, USAF did." He worked in an area that studies foreign technology, studying Russian items only, like the N1 project, AJAX plane project and the Mig Foxbat 25. Rutledge had previously worked on the Gemini project and the USAF remembered that, so they chose him later for Apollo 20. He says he was chosen, "because I was one of the rare pilots who didn't believe in God." For some reason, he said that little thing was the deciding factor making him an astronaut. Rutledge says that there were 300 people working on the mission for the USAF at Vandenberg, California. During this mission there was a Russian-American collaboration to make the mission happen.

Rutledge claims that the Apollo 18 mission was the Apollo-Soyuz project that took place in 1975. Apollo 19 however, had a loss of telemetry for some unknown reason, ending mission data sent back. It apparently hit a meteorite in orbit. Hmm...I believe the odds of a moving object hitting another moving object is far less likely than a meteorite hitting a stationary object, yet this could have been a warning sent from lunar inhabitants. Apollo 18's goal was to send a rover to explore the roof of a ship by climbing up Monaco hill and entering an opening made by those who explored the ship far before the USAF ever arrived.

When Rutledge was speaking about the alien ship and lunar city on the moon, his descriptions sent ones imagination whirling.

"We entered the two ships on the mission. One was a cigar shaped mother ship, the other a triangular-shaped ship. (Note: SCW confirmed that both ships do exist next to one another using digital enhancement described earlier on photos AS15-P-9630 and AS15-P-9625.) Most the mission fell on exploring the mother ship, which NASA believed crossed the universe at least 1.5 billion years ago. We witnessed many signs of biological life forms inside, like old remains of unusual plants in the section that apparently contains the engine for the craft. Another biological species came in the shape of triangular rocks that emit droplets of a yellow liquid that has medical properties. We found signs of extra-solar creatures. We discovered the remains of many tiny bodies measuring about 10 cm across. These tiny creatures were living in a complex network of glass tubes all along the interior of the ship. The most startling of all the discoveries was when we came upon two human like bodies, one of which was totally intact even after over a billion years on the ship." The one intact EBE was named by Leonov or himself (he was unsure) calling her Mona Lisa. He said the EBE (Extraterrestrial Biological Entity) was female and close to 1.65 meters tall. She had genitals, hair and six fingers. Possibly meaning their mathematics may have been based on a dozen. It was apparent that her function was to be a pilot. She had piloting devices attached to her fingers and eyes, yet had on no cloths. She had two thin cables that ran into her nose area, but note she had no nostrils. Leonov undid the devices over her eyes, causing secretions of bio liquid to shoot out and freeze from the mouth, nose, eyes and other parts of the body. Some areas of the body seemed in unusually good condition for its age. He stated, "As we told mission control, condition seemed not dead, not alive." A thin transparent coating protected her hair and skin. We didn't have a medical background so Leonov and I made a simple test. We decided to take off our bio equipment and attach it to the EBE so that it could send the telemetry back to Mission Control Medical experts. She is now on Earth and she is not dead, but I want to post other videos of her before I talk further about her. We also found a second body that appeared to be destroyed. We took the head on board. The color of its skin was a pastel blue or blue-grey. The skin had a bit of unusual inscription over the eyes and forehead. There was a strap without inscription around the head. The cockpit of cigar shaped mother ship was full of calligraphy and was made by lengthy hexagonal tubes.

The actual age of the spacecraft was 1.5 billion years old, which was confirmed during exploration. We discovered fragments of the original crust, anorthosite, spirals in feldspathoids, apparently came from the impact which formed the Izasak D Crater. "The density of the meteor impacts on the ship validated the age, also the little white impacts on the Monaco hill at the West of the ship," said Rutledge.

Rutledge continued saying the city that they discovered was actually named on Earth. It was decided that this city would be called Station One, but once astronauts got a close look at it, they realized that it was long since abandoned and turned to worthless piles of scrap, except one taller larger building they called the Cathedral. That one seemed intact. (The Cathedral is clearly shown on the far left of the video in detail on the video of the lunar city). There was a writing or calligraphy deeply pressed into every piece of metal. We believed the ship and the city were similar in age, but it is a very tiny part of the whole. On the rover our telephoto lens makes the alien structures look bigger than normal.

Rutledge said that he was surprised that no one ever asks him "why is it necessary to hide UFO's?" He says it's a question of economics. "All the currencies on Earth are based on the value of gold," he states. Most people don't know that gold is an exterritorial metal that is created when a star dies. When a star dies, its mass grows greatly, atoms get compressed and then the star explodes. Large amounts of gold get spread across young solar systems. That's why gold is a carbon free metal. This means that it is the most common substance in the universe.

I found this statement above to be bizarre at first, but then when reviewing the "Alien Spaceship on the Moon – Fly over before landing" video clip that he released, I noticed several reflections of yellow light from the surface. Two of these patches of what looked like 24k gold were seen from the LM (lunar module) as it flew over them. The video is at http://revver.com/video/624300/apollo-20-legacy-part-1-the-city/ and at http://www.youtube.com/watch?v=bf1G8JAGzt8 . The first patch of gold appears at the far left of the screen at 55 seconds into the clip. The second larger patch appears 1.16 seconds and on the far left of the screen. Two yellow areas that I had earlier assumed were reflections, turns out to actually be huge patches of gold color mineral. One circle-shaped patch appears to be several meters in diameter. The other is plainly three to four times larger than the first. Now upon closer inspection, the gold appears to be similar in color to unprocessed gold ore with a luminosity of 20-24k gold. When I showed the video to a geologist friend at a local university, he confirmed that it has a high probability of being gold according to its reflection of the suns light and its surface color.

Others with Evidence to suggest Rutledge's story is true:

1. It was reported in March of 1996 that NASA scientists and engineers who took part in the Moon and Mars exploration met in front of Washington

National Press Club to discuss their findings. They immediately announced that numerous man-made structures and objects were discovered on the lunar surface. The article goes on to say that there is a mass of evidence of abandoned cities. An unnamed mission control specialist stated, "Our guys observed ruins of the Lunar cities, transparent pyramids, domes, and God knows what else which are currently hidden deep inside NASA safes, and felt like Robinson Crusoe when he suddenly came across prints of bare feet on the sand of the desert island." Later they took back the statement and called it a mistake.

2. Sergeant Karl Wolfe was in a photo lab when his boss, Staff Sergeant Taylor, came over to him and told him they were having difficulty with some equipment on the base at the first Lunar Orbiter program. It was the same kind of equipment he was use to using, so he thought nothing of it. Wolfe was told that it was a NASA facility that he had gone to, but Wolfe remembers him saying NSA (possibly the NSA is in charge of Lunar artefacts). Wolfe said that he was called in to fix some photographic equipment, and his security clearance was raised due to him being the only person at the time able to fix such equipment. He saw a lot of activity in the hanger that day. He noticed that a lot of guest nametags were on individuals from foreign countries in this hanger. He was in the lab when an airman came up to him saying, "We discovered a base on the back side of the Moon." Wolfe was stunned of course. Then the airmen showed Wolfe some photographs that clearly showed geometric shapes, well organized

and well designed. This story confirms Williams Rutledge's details.

President John F. Kennedy wrote a memorandum for the CIA director about UFO Intelligence files on November 12, 1963, just eleven day before he was assassinated. The body of the memo stated, "As I had discussed with you previously, I have initiated (blacked out) and have instructed James Webb to develop a program with the Soviet Union in joint space and lunar exploration. It would be very helpful if you would have the high threat cases reviewed with the purpose of identification of bona-fide as opposed to classified CIA and USAF sources. It is important that we make a clear distinction between the known and unknown in the event the Soviets try to mistake our extended cooperation as a cover for intelligence gathering of their defence and space programs. When this data has been sorted out, I would like you to arrange a program of data sharing with NASA where unknowns are a factor. (Unknowns here may refer to knowledge of alien cities, structures and ships on the moon.) This will help NASA mission directors in their defensive responsibilities. I would like an interim report on the data review no later than February 1, 1964," signed "John F. Kennedy."

This memo clearly stated that the NASA was working on a specific space program with Russia! Perhaps it was the fact that JFK asked the CIA to reveal and share classified information with NASA officials that may have upset the CIA intensely, causing the CIA to feel that JFK has overstepped his bounds. In1994, Dr. Mitchell had stated to the St. Petersburg Times that there are many insiders within the US government that were studying the recovered bodies of aliens. That this specific agency has stopped briefing US presidents after John F. Kennedy. We may never learn the truth of who killed him, but we can assume that if no other presidents were advised of alien bodies, alien prisoners, alien captured or traded craft, and then we have to ask, what happened with JFK that made them stop telling US presidents? Second, when JFK asked the CIA in the memo to share classified info with NASA, did the CIA believe that it was almost like going public with it?

It is clear from the evidence and research, that William Rutledge is telling at least a partial truth. He does have a bit of sensitive NASA material and information about covert missions to the moon, but that information seems limited in what he shows. It is clearly genuine material and the abundance of debunkers hacking into his account and making debunking websites about Apollo 20, only make me believe these debunkers are being paid for all their hard work and have an agenda. I do however believe that he was an astronaut on one of these covert missions, due to the details of the Apollo 20 mission that *Retiredafb* or *ValValientThor* gave. This person is trying extremely hard to get his story out, yet he is not telling us much yet, so only time will tell if he is a true astronaut. Then again, how much information is enough to judge? Nevertheless, in his interview he mentions that the documents and video came from a crate guarded by a security guard that was suppose to watch over them until they were destroyed. Rutledge says his friend was shown these items by the guard and allowed to take what he wanted, latter sending some to Rutledge. I feel that the person calling himself William Rutledge could possibly be that security guard in the USAF. Also I find it hard to believe that NASA eggheads couldn't think up a better name for a covert mission, instead of continuing to use the Apollo names. By using the Apollo mission name, they would increase likelihood of being exposed. Rutledge former astronaut or impostor is a unique individual who deserves our attention.

Are you aware that there have been 59 lunar exploration missions (robotic and human) from January of 1959 to January of 1998? (Clementine Atlas of the Moon, page xix) These missions involve mostly the USSR and the USA, although one mission was from Japan called Muses A in January of 1990. This high number of moon missions is clear evidence that there is something up there to warrant the money and the risk…alien technology?

Surly his leaked information may have been the cause of Japan and China to launch lunar mapping satellites recently. Japan sent a satellite called SELENE to photo the lunar surface and reached lunar orbit on Oct. 18, 2007, at an altitude of 100km above the lunar surface. China's Chang-1 satellite was launched Oct. 24, 2007 set to orbit the moon for one year while sending back detailed photos. Two satellites launched from two countries in one week set for a similar mission. What do you suppose China and Japan hope to gain once they see the extremely highly detailed photographs of the lunar cities and ships? Only time will tell, but one thing is for sure, the truth cannot be hidden from us forever.

Chapter Six: UK Pilot Flies Past Two Mother Ship UFOs, 2007

Two extraordinary UFOs were seen in the United Kingdom on April 26, 2007. This was not the average sighting, not to say that most sightings are anywhere close to normal. The pilot estimated the UFOs to be almost a mile wide! Not one but two pilots filed reports with the Jersey airport air traffic controller, Paul Kelly. Mr. Kelly, age 31 confirmed that he received two reports at the same time from experienced commercial pilots who fly with Aurigny and Blue Island Aircraft. An official air-miss report on the sighting appeared in Pilot Magazine. Bowyers was literally shocked by the worldwide interest which has been in newspapers and talk shows around the globe. He took it all in stride until he saw a small article in the New Zealand Herald, the areas largest paper. Bowyers mother lives in Auckland and said she saw it in the paper and was very impressed with him.

16. Photo of UK UFOs. (At http://scwbook.blogspot.com/)

17. Photo of plane Bowyer was flying. (At http://scwbook.blogspot.com/)

Captain Bowyer says this about the sighting, "In all my years of experience, I have never seen anything like this and frankly, I would be perfectly happy not to ever again. It's a pretty chilling thing." He said that he was very scared and very happy once he got back on the ground and had a cup of tea to calm his nerves. He didn't have any fear of being ridiculed, because he felt that it was a true experience and that it was his duty as a pilot to report it.

The sighting occurred at around 3 pm. Fifty-year-old Captain Ray Bowyer was the Aurigny pilot who first spotted the cigar shaped UFOs. He says that he spotted the strange object during a flight from Southampton. He saw an incredibly bright yellow light that was ten miles off the coast, west of Alderney (island) when his plane was about thirty miles away from Alderney, and he was flying at an altitude of 4,000 feet. His passengers also saw the "cigar-shaped brilliant white light." It had graphite gray along the middle area that may have been the fuselage and was defiantly a solid object.

Bowyer states that as his plane got closer to the UFO, he used his binoculars with 10X magnification and saw, "It was a very sharp, thin yellow object with a green area. It was 2,000 feet up and stationary." Then he continued, "I thought it was about ten miles away, although I later realized it was about 40 miles from us. At first I thought it was the size of a Boeing 737," he said. He insists that it must have been much bigger because of the distance, and he now believes that it could have been closer to one mile wide. Bowyer has flown commercial planes for over twenty years. He says that the only reason he lost sight of the UFO was because as he was approaching for landing, he descended through a haze, which is very common in the area. After descending through the haze, he could no longer see the UFO anymore, but Bowyer believes it was certainly still there.

The second pilot who saw the UFO was flying through controlled air space, or military air space (Note: Possible base for UFO mother ships to dock in this Area, known by the local government). In other words, although he reported the exact same details concurring Bowyers story, the other pilots report and details cannot be allowed to air on public media without being seen as breaking the law. This is the reason the second pilot does not make appearances on television talk shows as Captain Bowyer has done. Bowyer has been on the Richard and Judy Show in the U.K. and many other news shows.

While continuing his approach to Guernsey, Bowyer also saw a second UFO similar in shape, but much further away to the west. "It was exactly the same, but looked smaller because it was further away. It was closer to Guernsey. I can't explain it. This was clearly visual for about nine minutes. I am certainly not saying that it was something of another world. All I am saying is that I have never seen anything like it before in all my years of flying."

Although numerous passengers saw the cigar-shaped crafts, only two of Captain Bowyer's passengers, John and Kate Russell also reported seeing the UFOs. John age 74 said, "I saw an orange light. It was like an elongated oval." One of the passengers allegedly was seen taking photos or video with a cell phone, but this cannot be confirmed and no passengers have come forward with photographs.

The Civilian Aviation Authority stated that a second craft, a Tri-Lander aircraft happened to be flying near Alderney, which also reported seeing the UFOs.

Strangely enough, Jersey radar equipment did not pick up the flying objects. This sighting is similar to the O'Hare sighting in that fact. It is clear that modern day radar is not equipped to catch our interstellar friends as they come near to or even on top of the radar. This could be the reason that the US government continually shuns at replacing old radar equipment of airports with new modern equipment made to find more than just flying metal objects. The US obviously wants to avoid such detailed evidence that would of course, leak to the public sooner or later.

The air traffic controller at Jersey airport said that the Blue Island pilot had given him a report about the flying objects, which confirmed Captain Bowyer's account.

Kelly said, "The description was very similar to Captain Bowyer's and they described it as being in exactly the same place. But they were looking at it from opposite sides." The other pilot was flying to Jersey airport, and as he flew past Sark, he saw an unusual object hovering behind him and to his left. This pilot for Blue Island Airlines also said that the huge UFO was 1,500 feet beneath his very plane! Both Captains said that they had seen the UFO at the exact same location. The Blue Island Captain said his plane was 3,500 feet up at the time of sighting the strange craft. Mr. Kelly stated, "If the object was stationary, our equipment would not have picked it up because the radar would have screened it out."

Although radar didn't pick up on the alien crafts, there were interesting traces of something that could have been evidence of the UFOs in the area. These traces are usually filtered out as anomalies by the radar, but when going back over the data, traces (spirit traces on both Guernsey and Jersey radar) were found that something large might have been in the sky that day. These traces existed on the radar for nearly fifty-five minutes, although Captain Bowyer says he saw the UFO for a total of twelve minutes. These spirit traces on the radar were not found until Captain Bowyer himself paid a visit to the radar station and looked into the recorded data with the help of one of the radar specialist. Mr. Kelly said that the traces on the radar of both UFOs do exist; yet the traces are not enough evidence to say that UFOs were actually in the area.

When Captain Bowyer was asked if the craft was something of this planet or something from outer space. Bowyer replied, "That's extremely difficult. I mean I think a few years ago, Dr. David Clark was interviewed about and asked the exact same thing, about four years ago today. He is in charge of a team that is looking into this at the moment. Uh…I have been asked this similar question by the press and my simple answer was, I don't think it was from around here."

Bowyer says that he as a pilot when flying has a heightened sense of awareness when flying. He says that these two UFOs were not the first UFO sighting that he has had, but that he has seen others in the past when flying. Now that he has come out and told the world what he has seen, many pilots often take him aside and say, "Hey, let me tell you what I have seen." Bowyer says that a lot of pilots fear ridicule and fear loosing their jobs, so they usually keep such sightings to themselves. Over fifty others who have seen UFOs around the islands have stopped him to chat about their experiences.

Captain Bowyer said that he was appalled at how the sighting was handled at the O'Hare airport in 2006. He stated this at a Press Club, "Despite many pilots and airport personnel witnessing the object hovering over the terminal, there was no investigation at all by the FAA. It appears a pressure might have been applied to crewmembers by their company not to discuss their incidents. I would have been shocked if they said to me that the CAA in the UK would obstruct an investigation or if the CAA had told me that what I had seen was something entirely different. But it seems that pilots in America are use to this sort of thing here. I would urge all fellow aircrew to report whatever they see as soon as possible and standup and be counted."

This UFO report never made much of an impact in American newspapers for some unknown reason, but then, the US news has been known to be working with the US government in keeping UFO sighting information to a minimum, except Fox News, which valiantly follows their own rules.

There appears to have been a lot of sightings, most of which go unreported around the triangle area of Alderney, Guernsey and Jersey. Bowyer said that its extremely strange that so many sightings would come from one little area over the ocean. The fact that so many sightings could be made of hovering UFOs in the area can lead me to only one conclusion; there must be a base on the ocean floor, somewhere in that area. Hovering would indicate that it was resting after it came from above or below. Also remember that while these craft can be incredibly large, they do have technology to conceal the ship from a long distance. Usually this only works if the viewer is looking from below the craft, passengers on aircraft in flight seem to see these phenomenon more easily than those on the ground. Also a sunset or sunrise can affect the stealth abilities of a UFO. Many sightings seem to take place at a moment in time when the sun light is changing, allowing a different spectrum of light to bounce off the surface of the UFO, thereby making it visible to the human eye, or so evidence of past sightings point in that general direction. For instance the O'Hare sighting took place right before sunset. Bowyer himself seems to be an extremely likeable and trustworthy person, which is helpful in rallying support around the world for his investigation. The French authorities however refuse to comment on the UFO sightings and refuse to allow the pilot of the Blue Island plane to comment either. The fact that these two mile long UFOs were flying over French Military restricted waters may indicate that the French have some underwater facilities in that area that they do not want the public to discover. Possibly using a satellite map of the bottom of the ocean would be helpful. Remember, if these UFO crafts can travel through space, then it would have no problem traveling under water.

Chapter Seven: Three UFO Sighting at Area-S4 in 2007.

The place that has become linked to UFOs around America is often called by most by the name of Area-51, but is also goes by the names Groom Lake, Dreamland, Watertown Strip and Homey Airport. The granite hard secrecy of which the US Government barely acknowledges surrounding the base called Area-51, has in itself created numerous conspiracy theories and UFO history since the Roswell, NM crash in July 7, 1947, of which, its remains, including two aliens, one alive and one dead, were brought to Area-51.

18. Photo of Roswell Newspaper of 1947. (At http://scwbook.blogspot.com/)

19. Drawing of Lazar's UFO. (At http://scwbook.blogspot.com/)

20. Photo of runway at Area S4. (At http://scwbook.blogspot.com/)

21. Photo of triangle UFO on hill. (At http://scwbook.blogspot.com/)

22. Photo of small round UFO. (At http://scwbook.blogspot.com/)

23. Photo of large UFO under hanger. (At http://
scwbook.blogspot.com/)

In July of 2007, I found three UFO's when I was exploring a
lead about Bob Lazar. He claimed that he was once an employee
at Area S4 and that not only did he work on a flying saucer that
was not made by humans, but he says that nine other saucers
were kept in a hanger at the edge of a mountain near Groom
Lake. I wanted to find Lazar's Hanger, so I began instigating
this not because of his story, but because I noticed a lot of people
were trying to show him as a hoax and that could mean only
one thing, people are being employed to debunk him. Well, this
caught my attention and I decided to investigate the area that
he described, but since people are not allowed to enter within
five miles of the outer gates, I looked for a more simplistic way.
I used Google Earth. This program was free to download and it
gives you a satellite view of any location on the planet.

It took only the time to type Area 51, into the search before
I found myself staring at the infamous base. I was amazed that
they allowed such details of the base to be shown, but soon I
noticed some rectangular areas that were poorly covered up
with a photo program, making them appear like sand. What
was I looking for? Easy, anything out of the ordinary, especially
anything saucer shaped (classic right?). I know, beware of
circular water towers and radar towers. Those are easy to
identify, if not from their structure, then from their shadows
they make. I soon found a few F-16's sitting on the runway and
two other larger jets, but nothing unusual. Then I remembered
that it was near Groom Lake. That was a dry lakebed that now
appears white. I zoomed out from Area 51 so that I could see
the area better and found that on the left side of Area 51, close
to seven miles traveling southwest, was a dry lakebed, but this
is only the second lake bed. This one is often confused with the
third dry lakebed. The third one is where, with Google Earth,

you will find the Three UFO shaped craft. Area S4 is located roughly fifteen miles South-West of Area 51.

First UFO Sighting: Steps to follow for finding 108-diameter saucer:

1. Download Google Earth.

2. Type "Area 51," in the search box.

3. Zoom out until you see two more dry lakebeds on the left of Area 51.

4. Zoom in on the third lakebed with your mouse. Now you are very close.

5. Zoom in at the Southern most tip of the lakebed.

6. You should see a short runway with a dirt circle on one end.

7. Zoom in on the circle where you will see planes and hangers.

8. Zoom in on the largest gray hanger. It's five times larger than others.

9. Half under one hanger you will clearly see a UFO of 108 feet in diameter. Use Google ruler to take some measurements of its size, height from ground and so on.

10. Coordinates are: 36,55'35.72" N 116,00'25.33" W

11. Google photo date placed for this location is Dec. 2, 2006.

Second UFO Sighting: Steps to follow for finding the 42-foot in diameter saucer while in flight:

1. Start at the 108 UFO at the South side of the same lake as above.

2. Zoom out so that you can see more of the map above you.

3. Find the purple place dot above marked, "Nevada Test Site."

4. Move to the purple dot.

5. Zoom in a bit until you see a second purple dot called, "Sedan (Nuclear Test)."

6. Move to that dot. Then look above it on the map.

7. Look for the yellow dot marked, "US Nuke: Orkney."

8. Zoom in on this spot, because 400 feet to the left is the saucer.

9. Saucer is dull gray. Size is about 42 feet in diameter using Google Ruler. Note that its shadow below is separate from the UFO, meaning its in flight and you can measure its height from the ground with the Google Ruler.

10. Coordinates are: 37,11'55.01" N 116,03'19.93" W

11. Google photo date placed this location is Dec. 2, 2006.

Third UFO Sighting: Steps to follow for finding the "V-shaped" 64.66-foot total wingspan in flight craft:

1. Start at the Area 51 icon at the base or type AREA 51 into the Google search box again and it will take you back there fast.

2. Use Google ruler and go 3.14 kilometers or 1.95 miles Southwest of Area 51. Here you will see many small mountain peeks.

3. Upon one mountain peek is a flying V-shaped craft, which leaves a diamond shaped shadow on the ground directly below it.

4. Its coordinates on Google Earth are 37°13'02.94"N and 115°50'49.35"W

5. View this on a digital monitor, while looking through a digital camera for a more detailed view. Also rotate the Google map from North to South using the compass to view the UFO craft properly, so you don't view it upside down, because of the satellite angle.

6. Google photo date recorded this location on Nov. 14, 2006.

Now that you found the first larger UFO, lets talk about what you see. The 108-foot saucer appears to be dull gray metallic in nature. It sits with close to 60% of itself being revealed from under the hanger. There is no reason to place a fuel tank or water tower under a hanger, so we can assume that the USAF parks aircraft in their hangers attached to airfields. That is our first assumption.

Second, notice the tall device connecting to the south side of the ship. It seems to be attached to the ship at a fixed point and four legs supports this unusual machine. From the attached point to the ground it is twenty-five feet. I know this because of both Google Ruler and the shadow that this machine leaves below it on the ground. There appears to be three 75-foot metallic color tubes attached to one side and leading to a white flat area with parked equipment on top. Also there are two 75-foot long tubes of similar size coming out of the saucer and going to the square white area above it on the map. We can assume that the tubes attached to the saucer have a function, maybe cooling, since the craft is exposed to the sunlight in the desert and it appears to be in a maintenance area.

Know this, other UFO's that people found around the world on Google Earth, got mass media when the people who found them, photographed them and put them on the internet for all to view. Unfortunately, Google Earth also saw these and deleted those objects out of their satellite maps within six months of their release. I on the other hand will not release this data until the release of this book, so that only those who read this book will have a better chance of still seeing the saucers at Area S4, without risk of Google deleting them. To find the UFO's that other Google users have found, simply open up your web browser and push photos, then type, "Google, UFO." Numerous photos of the circular shaped metallic objects have been found using Google. They have photos of them flying over forests, neighborhoods and parks all over the world. Most of these UFOs look the same; chrome like spheres.

Who is Bob Lazar and how does he know about UFO's you ask? Lazar has degrees in Physics and Electronics Technology and says that he has attended Cal Tech and MIT. In the 1980's he worked at Los Alamos Labs in New Mexico. Then in 1989 he was employed at a desert test area called Area S4. Area S4 is about 15 to 18 miles South West of Area 51. It is 125 miles North of Las Vegas. Area 51 has become notorious for the secret tests of military craft such as the U-2 and Stealth Fighter.

Bob Lazar believes that there are no alien craft being tested at Area 51, but instead the UFOs are at Area S4. His job at Area S4 was to assist in the back engineering of the propulsion system of the alien saucers. Back engineering is to analyze something that already exists and figure out how its works, essentially by taking it apart or close study. One of the Area S4 briefings that taught Lazar about the alien ships and other technology, also revealed a startling piece of information to him he didn't know, that humans were products of genetic manipulation by the aliens themselves.

Lazars saucer that he worked on has to be one of the most researched and well-known UFOs in the world of UFOlogy. It is called the Sports Model, and Bob Lazar worked frequently on it at the S4 sector, near Area 51. The dimensions of the sports model UFO are approximately 16 feet tall and 52 feet nine inches in diameter. The outer covering of the disc is made of a metal and its color is similar to unpolished stainless steel. When the saucer is not energized the craft will rest on its belly. The entrance to the craft is found on the upper half of the disc with the bottom portion of the door wrapping around the center lip of the discs edge. This was just one of nine discs kept at the S4 site. The saucer itself was actually capable of not just space flight but traveling forwards and backwards in time. He says when you create any artificial gravitational field, you technically move in your own time. So you do slip forward when you create your own intense gravitational field, but he says that has yet to have been tested at his site.

Lazar describes the disc as being divided into three different sections. The lower level is the location of the three powerful gravity amplifies and their wave-guides. These components are the main part of the propulsion system that is used to amplify and focus the gravity A waves.

He states that the reactor is found above the three gravity amplifiers on the center level. The reactor is centered between all three amplifiers. The reactor is powered by an element that up till today, was only found in meteorites, element 115. This element has never been found on earth, but was once found in a meteorite that fell in South Africa. Element 115 is the source of the gravity A waves, which is amplified for space-time distortion and travel.

The center level of the disc has the control panels and seats, but Lazar said they were both too low to the floor and too small to be functional for the average adult human. The walls were divided into archways. A unique thing happens when the disc becomes energized, one of these archways becomes transparent and you can actually see the area outside the ship. The outer wall of the disc between the archways became a window of sorts. When Lazar was observing a demonstration of the interior of the disc as it was energized, he noticed something incredible, which is only similar to our LED technology of today. A form of writing began to appear on the right side of the clear window, which was unlike any alphabetic, scientific, or mathematical symbols that he had ever seen. The screen with the alien words and the clear window had a thin white bar that had a function much as our scroll up or down key works in our Word Perfect or web browser programs, allowing us to move it up or down to view more details. He was never allowed to view the third level, so Lazar was unable to give any specifics as to what the porthole areas at the top of the disc actually were, but one thing is for sure, they were not portholes (possibly a weapons area, which would explain the window like openings at the top that would allow a particle beam to shoot out.). All three seats in the craft face the same screen. The words on the screen were composed of small circles, and ellipses, much how cursive writing curves beautifully, but not connected like cursive words. Lazar was not told what the writing meant or where it came from, so his only moment of seeing it was when the technicians switched it on from a hole in the floor of the center of level. The hole was made when one of the gravity amplifiers was cut out for analyzing. From the moment the technician started the mechanism, to the moment Lazar viewed the words on the clear screen took about thirty seconds (a delay of 30 seconds before the screen started). The ship had only a single weapon on it. It was a direct energy weapon. Unlike a laser beam this weapon was probably a particle weapon or neutron weapon. Also Lazar said that the EMP generator in the system of the craft could possibly cause approaching jets or systems to malfunction. The craft themselves make a slight hiss when taking off and can be heard up to a certain altitude, and it sounds like a high voltage hiss. He said that the interior and exterior of the craft appeared as one piece, as if it came from a wax mold. There were not nuts or bolts or wires connecting anything. There were no lights on the control console.

The element 115 fuel sits on top of the conical shaped fuel housing. The element 115 was machined into long isosceles triangle shaped wedges with a conical point at the angle adjacent to the two equal length sides of the wedge for fitting into the reactor. A Los Alamos facility machined the 115 into these triangle shapes and then shipped them to Area S4. Lazar says that he had a contact in Los Alamos National Laboratory in New Mexico that took part in the machining of the element and gave Lazar one of them. At Los Alamos they believed it was for some futuristic armor, they had no idea what it was really meant for use as. The wedge that they created was about the size of a 50-cent coin, but was suppose to last the flying saucer for at least the next twenty to thirty years, depending on how much use the craft got. This tiny mineral called element 115 is not only the key to starting and running the many different types of alien craft that Area S4 has, but it is also the key to unlocking free energy, freeing vehicles from their dependence on fossil fuels, allowing the planet to heal the ozone naturally. Imagine not just cars and aircraft, but everything everywhere running off this tiny coin size mineral that lasts for decades before it's used up. Sounds like the future to me.

Overall I feel that Bob Lazar is telling the truth about himself and the work he has done at area S4. The three UFO's that I showed you in this chapter came from using Google Earth to prove that there are in fact, flying saucers being flown and housed in hangers at Area S4. Perhaps it was one of these test pilots that arrogantly though that he would not be detected over O'Hare Airport in 2006. Lazar also said that the three anti gravity generators also created a field around the saucer that made it appear fuzzy or unclear to some degree. He also mentioned that this field around the saucer might appear illuminated at night due to its high intensity of micro-like waves. This would explain a lot of photographs in the past that have had UFOs that appeared slightly fuzzy. If Lazars statements on the field around the ship are correct, it makes all those past blurred photographs now appear real. The fact that in 1995, the Department of Interior took 3,972 more acres (around Area 51 & Area S4) of Bureau of Land Management land, from public access, creating a extreme security buffer zone to prevent any civilian from observing military activities at Groom Lake, or in other words, in 1995 they started teaching pilots to fly these alien crafts both in and outside the atmosphere and needed a buffer zone to prevent the crafts from being seen when at high altitudes. No, this is not frantic guessing, but tying in the facts. One fact that backs this up is the UFO crash in Needles, California of 2008, where two Janet planes (exclusively used in Area-51 to transport non-military personnel to work.) came in the middle of the night to transport scientists to examine the crash. Strange the planes landed hours before the crash took place, indicating it may have been an Area S4 pilot that crash-landed a blue glowing ship or an expected alien visitor. Check out the 2008 Needles UFO crash chapter that was 178.6 miles Southeast of Area-51.

Chapter Eight: Alien Evidence on Lunar Surface found in 2007.

If you want to see the structures for yourself on the moon, then please pick up your digital cameras or DVD cameras right now and go to find some NASA photographs, enlarge them by 5X, then start scanning them on a flat computer screen while looking through your digital camera. Note, holding the camera at a slight angle (100-120 degrees) up will reveal hidden items that will first appear as ghostly clouds over the ground, but will then reveal to be structures. Digital begets digital, meaning that the digital photographs on a digital flat computer screen, seen through your digital camera will cause the once blurry 5X photo to re-digitalize and become focus again, sometimes gaining 100% focus. It may take a half hour or more to find your first structure, but once you gain in experience, it will get easier and easier till, like me, it takes mere minutes if I use a photograph from the Apollo Image Atlas at http://www.lpi.usra.edu/resources/apollo/.

24. Photo of dark moon structures (flat black in color). (At http://scwbook.blogspot.com/)

25. Photo of ancient triangle craft in the left side (Apollo 20, Deporte Crater). (At http://scwbook.blogspot.com/)

26. Photo of lion head structure in upper left. (At http://scwbook.blogspot.com/)

Then from there, go directly to the word 'Panoramic.' These are the easiest NASA images to enlarge and see structures with windows or doors. Sometimes the structures are even in the shape of an alien head. One building I found was in the shape of three aliens faces, each different making me believe it was a commemorative building of different species working together. Then go to which Apollo mission you want to see more. Apollo 15,16,17 are all available. Apollo 15 has 1529 images. Apollo 16 has 1,435 images. Apollo 17 has 1581 images available for you. Then you must follow these simple steps:

1. Pick a photo, click on it. Then push your right button on your mouse to "save as," then choose to save it to your "desktop."

2. Once on desktop, click the picture using your right mouse button again, choose "open with" then choose "Windows Picture and Fax Viewer."

3. Once open, click on the magnifying glass + icon then click on the photo to make it larger. Enlarge it 3-5 times its original size (300-500%).

4. Once finished, start scanning over it with your digital camera. I found that my Sony Cyber-shot with Full HD still images and 10.1 Mega Pixels gives great photos with no digital lines ruining them. Other older digital cameras will work, however, fuzzy lines may appear. If the lines do appear, don't give up; take the same photo from different angles or distances away from the flat screen, one will come out clear if you keep trying.

5. Often, especially in the panoramas of Apollo Image Atlas, there are hidden structures, that will be revealed if you move your camera angle sweeping back and forth from a 0 degree angle to 30 degree angle, actually looking upwards at the screen. This reveals much more than you would believe, so the only way for you to understand is to try it and see for yourself. I'm not sure why this works, but it does. The first time you try this successfully, it will take your breath away.

This technique will work on most digital astronomical photographs, so now you can start exploring the universe in the comforts of your own home. Explore the moon, Mars, IO (my favorite) and all the other planets around.

I discovered the moon structures first in photograph AS15-P-9625 & AS15-P-9630 when researching William Rutledge story about his Apollo 20 moon mission. The research took me to the most obvious evidence available, NASA photo archives. This is where I discovered the first and most compelling evidence that he was telling the truth about covert missions to the moon. I found the panoramic photo with the cigar shaped mother ship that Rutledge described. It was close to Deporte crater just as he said, yet I saw a lot of details in the photograph that he did not mention. I saw a triangle craft next to the cigar ship, but this craft was covered in a few feet of dust. There were several rectangle cave openings going into the triangle craft as if someone dug inside to investigate the craft. Rutledge did describe how when he was on the moon he also investigated a triangle craft next to the cigar ship. This must be the craft. He also talked about the holes made to enter the triangle craft that were made far before the US astronauts got there.

My research on him soon took me to the best and highest quality photo of Deporte that I could find at this link: http://www.lpi.usra.edu/resources/apollo/frame/?AS15-P-9630   .

The panoramic photo here was literally out of this world and also AS15-P-9625 shows the area more lighted, but making structures less visible compared to photo AS15-P-9630. Yes you heard right. There are clear and obvious structures that appear to be buildings and others that look like machinery. I even saw several ships. The ships varied in size from a hundred meters to half a kilometer in length. Some were parked in cave like crevices along craters with only their nose and cockpit windows sticking out. The cockpit and nose looked identical to that of the US stealth fighter jets.

When looking over image AS15-P-9630 with the naked eye, I happened to fall upon clear and obvious structures. A pipeline (possible transportation tunnel) that cut through the top to the bottom left of the photo. That pipeline must have been over a hundred meter wide and hundreds of miles long I assume, since both ends are cut off by the photo's edge (and this tunnel is seen to continue its way through many other photos). I thought that this pipeline might be a taped and glued area of the photo, but then I noticed at its bottom, two smaller pipes clearly overlapped the larger one in an X formation. I saw some tower-like white objects and a white triangle in the far left shadows of the photo. I felt excited and began to look closer. I plainly saw a dull silver building in the shadows. It's right edges and windows standing out from its shadowy background. At that moment I decided to copy the photo to my computer and enlarge it in hopes to get a much better view. I found that 500% enlargement or 5X was the best for viewing it. Then I added a tint of yellow to help details stand out. I could see structures that I had not seen before, yet they were blurred not from enlarging, but from NASA using a simple photo-altering program of some sorts.

In the midst of my despair, I picked up my digital camera to see if I could take a picture of the blurry building in order to get it captured in a larger size (megapixel) photo since the original photo was only 150 kilobytes'. Never in my wildest dreams have I seen such a site. The simple Panasonic digital camera that I used had actually re-digitalized the blurred pixels creating a much clearer image! It appeared close to 70-90% what the original photograph might have been.

This was not the first time that I the writer found NASA images to contain building like structures. I also remember as a junior high school student thumbing through the Clementine Moon Atlas during numerous lunch times, which consists of endless pages of amazingly detailed lunar surface photos. Many of these photos would contain a few blurred spots even though all the things around them were in extraordinary focus. When I looked closer, I often saw areas of buildings sticking from parts of the blur. I showed this to my high school science teacher and he was amazed at what he saw, but then said it may have been a misprint of the image. True, it may have been, but as an adult I went back and looked again. The parts sticking out from behind the blur spots were sharply focused and contained right angles or angles that could only have been made by a species with outstanding intellect unlike our own.

My high school experience was not my first discovery in NASA photographs. When I was in first grade, I saw a newly published book on Mars sitting in the elementary library on a podium. I thumbed through it looking for objects on the planet that could be intelligent beings, then I saw it for the first time; the face on Mars. I was surprised and showed all the students in the library including the librarian. Our teacher was so excited that she told other teachers to come over to see it. That was thirty-four years ago in San Jose, California. I don't know why I have been curious about this or why I continued to be through the years, but there is one thing that I am confident of, and that is I have to help others around the world become aware of my discoveries. The human race cannot continue to live in a pre-Galileo mentality. We have to make our species aware of this if we want humanity to advance as fast as our technology is distinctly advancing.

Science dictates to us that the Moon is Earths only natural satellite and is the fifth largest moon in the Solar System. The distance from the center of Earth to the center of the Moon is approximately 384,403 km. That means it is at the distance equal to thirty Earth's put along side one another. The moons diameter is 3,474 km, that is about one-quarter that of Earths diameter. The Earth orbit and rotation never allows one side of the moon to reveal itself to Earth. Stop there for a moment. The moon dark side is never visible from Earth? The secrets of the dark side…were accidently revealed in a conversation with Neil Armstrong long ago.

According to the Neil Armstrong (NASA Astronaut), the aliens have a base on the moon and told us in no uncertain terms, to stay away. Apparently both Neil Armstrong and Buzz Aldrin saw UFO's shortly after their history making landing on the moon during Apollo 11 on July 21, 1969. At a NASA symposium Armstrong and a professor had this conversation:

> Professor: "What really happened out there with Apollo 11?"
>
> Armstrong: "It was incredible, of course we had always known there was the possibility, the fact is, we were warned off! (by aliens). There was never any question then of a space station or a moon city."
>
> Professor: "How do you mean, warned off?"
>
> Armstrong: "I can't go into details, except to say that their ships were far superior to ours both in size and technology. Boy, were they big! And menacing! No, there is no question of a space station."
>
> Professor: "But NASA had other missions after Apollo 11?"

Armstrong: "Naturally. NASA was committed at that time, and couldn't risk panic on Earth. But it really was a quick scoop and back again." (Note: he said scoop, as if the mission was to retrieve an alien ship.)

Dr. Vladimir Azhazha said: "Neil Armstrong relayed the message to Mission Control that two large, mysterious objects were watching them after having landed near the moon module. But this message was never heard by the public, because NASA censored it."

According to a Dr. Aleksandr Kasantsev, astronaut Buzz Aldrin recorded the UFOs while he was inside the module using color movie film, and continued filming them after he and Armstrong went outside. Armstrong later confirmed that the story was true, but refused to go into further detail, beyond admitting that the CIA was behind the cover-up.

Astronaut Buzz Aldrin in an interview with the Science Channel stated this on camera about seeing and being nervous to report the UFOs they saw:

"Well obviously the three of us were not going to blurt out, hey Houston we have something moving along side of us and we don't know what it is. You know, can you tell us what it is? We weren't about to do that. Uh…because we know that those transmissions would be heard by all sorts of people and who knows what someone may have demanded, asking us to turn back because of aliens or whatever the reason is. So we didn't do that."

Instead Buzz Aldrin said that he cautiously asked Houston how far away was the S4B ejected section, but Houston still didn't understand the astronauts coded words, until latter when Apollo 11 astronauts made it to the moon.

Otto Binder, a former NASA employee said unnamed radio hams with their own VHF receiving facilities that bypassed NASA's broadcasting outlets picked up the following conversation:

> NASA: "What's there? Mission Control calling Apollo 11…"
>
> Apollo 11: "These "babies" are huge, Sir! Enormous! Oh my god! You wouldn't believe it! I'm telling you there are other spacecraft out there, lined up on the far side of the crater edge! They're on the Moon watching us!"

If the warning away from the lunar surface by the aliens really did take place, then that would explain why the moon landings stopped and why we have not tried to build a moon base. It does appear that it would be better and easier than an orbiting space station with no access to any raw materials or supplies. It sounds like NASA is taking the warning seriously. It was a warning that they wont be able to hide for long. This may be why NASA allegedly cancelled the Apollo missions at Apollo 17, saying their budget couldn't cope with it. William Rutledge says he himself was an astronaut on the covert Apollo 20 mission that worked with Russian crewmembers. It wouldn't be hard to send a craft into space while people are watching if people thought it was only a spy satellite being sent up. Rockets taking satellites into orbit around the Earth have become commonplace and mostly ignored. Covert missions to the moon sound plausible when we research all the data that confirms alien life on the lunar surface.

Milton Cooper, a Naval Intelligence Officer said that not only does the alien moon base exist, but also the U.S. Naval Intelligence Community calls the alien moon base, "Luna." They say that there is a huge mining operation going on there and that the aliens keep their huge mother ships there while they make trips to Earth in smaller craft.

Astronaut Milton Cooper said this city called Luna is located on the far side (dark side) of the moon. It was seen and filmed by Apollo astronauts. A base, a mining operation using giant machinery, and the huge alien crafts described in sighting reports as mother ships, exist there.

Some facts that point to a hollow moon:

A leading scientist at NASA, Dr. Gordon McDonald published a report in the Astronautics Magazine, July 1962. In it he states that according to an analysis of the Moons motion, it appears that the moon is undoubtedly hollow. If the astronomical data is reduced, it is found that the data requires that the interior of the moon be less dense than the outer parts. The moon would therefore show that it is more like a hollow than a homogenous sphere.

When Apollo 12 landed on the moon on November 15, 1969, crewmen Charles Conrad, Dick Gordon and Allan Bean had a UFO sighting. They reported to Mission Control at Houston that they had sighted two bogeys (UFOs). They also set up seismic equipment around the moons surface to read any earthquake the moon might have. When the crew had finished and were back aboard the Command Module, "Yankee Clipper," they took off back to Earth. Soon after launch, they discarded the assent area of the craft sending it to fall back to the moons surface. The impact of the discarded stage (40 miles for the Apollo landing site), sent an impact that was so loud that it was measured on the seismic equipment back on the surface, which felt its ringing for over an hour! This unusual phenomenon was repeated with Apollo 13 (intentionally commanding the third stage to impact the moon), with even more startling results. Seismic instruments recorded that the reverberations that lasted for three hours and twenty minutes and traveled to a depth of twenty-five miles, forcing the conclusion that the moon has either a light core or hollow core.

The moon may be far older than NASA had expected. Maybe even older than the Earth or the Sun. The Earth is estimated to be 4.6 billion years old, yet some moon rocks were dated to be 5.3 billion years old. The moon dust that is spread throughout the surface of the moon is estimated to be even older at 6 billion years. Also the rocks and the dust are both different in composition, meaning the dust didn't come from the decay of the rocks, but from somewhere else.

It seems strange that the moon is just the right distance, tied with the exact diameter, to completely cover the sun during an eclipse?

Isaac Asimov says: "There is no astronomical reason why the moon and the sun fit so well. It is the sheerest of coincidences, and only the Earth among all the planets is blessed in this fashion."

Ancient Greek authors Aristotle and Plutarch, and Roman Authors Apolllonius Rhodius and Ovid all described in their writing a group of people called the Proselenes, who lived in the central mountainous region of Greece called Arcadia. The Proselenes claimed title to this land for the reason that their forefathers were there "before there was a moon in the heavens." Their assertion is backed up by symbols on the wall of the courtyard of Kalasasaya, near the city of Tiahuanaco, Bolivia. This record shows that the moon came into orbit around the Earth between 11,500 and 13,000 years ago. Long before recorded history.

On February 28th 2008, US space scientist announced that a highly detailed map of the moons South pole area was being created. The high-resolution images are said to be invaluable in planning future lunar missions. Researchers at the US space agency's Jet Propulsion Laboratory (JPL) in Pasadena, California stated: "With this data we can see terrain features as small as a house without even leaving the office." Here we see that the JPL used the word "house," which could very well be a

Freudian slip, leaking data that they really want to talk about, but can't or won't.

It is more than feasible that the reason the moon never shows one side to Earth is because there is an alien presence on it or within it, keeping widely used areas hidden, but the lighted side also has some structures that can be seen using my method. As outrageous as the moon is a alien ship theory is, all of the evidence points to the clear fact that the moon is indeed hollow and being used by aliens and was brought here eons ago by an intelligent species. This is the only theory that is supported by all the data that I present and I could find no data that could disprove it. NASA tries to hide it for obvious reasons. First, they do not want to cause mass hysteria around the globe. Second, they do not want to compete with other countries to get the alien technology on the moon. Third, NASA is scared that another country, lets say China (currently sending their second moon mapping satellite) may achieve a man landing on the moon soon. Let's hope that those countries that explore the moon, act so with caution and without weaponry on board their ships.

On 7-21-2009, US Astronauts met with President Obama in the White house to discuss of course...the moon. Charles Bolden, the new head of NASA as well as Armstrong, Aldrin, and Michael Collins all visited Obama together on the 40th anniversary of man's first moon landing. Did the astronauts relay to the president what they had seen on the moon, alerting Obama that aliens are real? It would be highly unlikely that the astronauts of Apollo 11 could hold back such information from the new president, but rather hope to educate him with their knowledge. Recently in February of 2010, all moon mission programs were cancelled and permanently ended by the Obama administration. Why would the US government hide such important information such as aliens existing on an artificial moon that orbits Earth? Lets explain this in the simplest terms. Do you as a parent always tell you children every single little detail of everything that has happened to you in your life? No, you do not, nor do you have the time or the patience to take on such an endeavor. Instead you tell them what you want to tell them in order to keep their minds at ease so that they can continue to focus on their own lives. The parent is the US government, and we are the children they hide the secrets from. That's it in a nutshell. Maybe you should tell your children more huh?

Chapter Nine: January 24, 2007, Large UFO over Winslow & Leupp, Arizona.

Numerous residents of Leupp, Arizona were startled when a UFO was spotted flying over their city, which began at 7:40 pm Wednesday and continued till 9 pm that night. Two eyewitnesses to the event came out to bravely reveal what they had seen that evening. Sean and Deanna Dover were driving home from Flagstaff Arizona heading to Leupp when Deanna spotted an odd-looking craft hovering far above their Honda Accord.

27. Photo of Phoenix Lights (also at Winslow & Leupp in 1997). (At http://scwbook.blogspot.com/).

Deanna, age 20 stated: "I saw a bright light and told my brother to watch it." She hesitated for a moment. "Then it disappeared, then reappeared. I didn't see it as much as he did, because I had to concentrate on driving."

Her brother Sean, a senior at Sinagua High School nodded. He states: "We were about 10 miles out of Leupp and my sister said she saw something. It had a circle around it and was about one and a half miles above us. It had three lights and was triangular shape. We kept on driving and when we reached Leupp; we saw it had four lights.

They then watched as the UFO slowly flew over the Church of the Nazarene and the Leupp Public School.

"When it reached the edge of town, it went into stealth mode and had only one dim light," Sean said.

The two continued to drive through Leupp and rushed home to tell their parents what they had seen that evening.

Sean said, "We went inside and got my dad's night vision goggles." He explained that his father is a Navajo Nation Police Ranger and has equipment that has a special device on it to allow the user to see clearly in the dark. At this point they all walked outside to see if they could watch the strange craft.

Sean stated, "Then two jets intercepted it in the air. They came from the southwest. Then it headed east. Eight minutes later it lost them and circled back to Leupp. When it got here it started blinking its lights."

From Sean's description of the event up to this point, it appears the UFO although chased out of the area, returned to finish what it had started. Perhaps to communicate with a hidden underground alien facility or to scan the area for a mineral or lost probe. Whatever it was doing…the fact remains, the technologically advanced craft ditched two military jets then came back to its original position over Leupp. This indicates it had a purpose that was yet unfinished.

Sean was totally confident in identifying the object as a UFO and not mistaking it for some other form of aircraft.

He stated, "It was traveling too fast to be an airplane and not emitting any noises that a helicopter or plane would have made flying that high."

Note, that the fact that it was silent and had no visible means of propulsion and no identifying sound of propulsion is noticed during most UFO sightings. This allows the craft to sneak up and pass by homes, neighborhoods and communities without raising awareness, unless you are one of the lucky ones to be looking up at the sky at the very moment it passes.

As Sean's family watched the UFO fly around Leupp, flashing its lights in almost a signaling pattern. Sean's mother Daisy called her friend and coworker Denise Fredricks who went outside and also saw the UFO. Denise is a fifth grade teacher at Leupp Elementary School.

Denise states: "I was coming home from my son's basketball game in Dilkon. When I came inside the house and my husband said, 'Daisy just called and said there was something in the sky,' then I saw a triangular thing go overhead."

The Fredericks and Daisy Dover then met up at the center of Leupp near the gas station and watched the events with about 30 other UFO eyewitnesses.

Denise said, "There were two other planes and another aircraft that it looked like was refueling them." Then she spoke of the UFO still in her line of sight. "It was flying straight and then it turned. It went toward Bird Springs and then it turned toward Tolani Lake and then came back this way."

The two jets refueling while a triangular UFO flies around in their line of sight, may indicate that the UFO may have actually been a government craft or they saw it as a non threatening craft and the jets wanted to show their disapproval of the UFO in area. Jets rarely if ever refuel while trying to intercept UFOs. The fact that these two jet were refueling from another aircraft tells us they have been trailing the UFO for a long time and that they felt it was no threat whatsoever, allowing them time to refuel. This is very odd for the USAF. They consider any bogey that enters their airspace to be a potential threat. To refuel in front of the UFO is almost unheard of in UFO sightings.

All of the eyewitnesses to the UFO described the craft as a triangular shape with three or four tiers or levels to it. On the bottom of the craft was a sphere with a pulsating light. They said the UFO circled the area 15 times.

Denise and Daisy said that the UFO emitted a yellowish light and was approximately twice the size of the Leupp Elementary School Gymnasium.

As the UFO embarked on its 15 loop around Leupp, the craft's lights suddenly went dark. At this point sudden and rapid illumination shot across its edges in a pattern formation as if to send a signal to someone nearby. Just as sudden as it started, the flashing stopped and the craft began to make its way towards Winslow in the southeast.

The eyewitnesses said that several of them were carrying cameras and did take photographs of the craft. However, because of the darkness and the extreme distance of the ground and the aircraft, capturing the UFO clearly was impossible. Photos only revealed the UFOs lights and could not make out the form of the craft itself.

Every eyewitness said the events were unique due to the close proximity of the UFO to the ground, not to mention that the craft was seen flying alone.

This UFO sighting was not the first for the area, but rather one of many in history of Leupp. One witness, Sean stated: "It was really interesting to watch. I believe that it really was a UFO because of my family's history. I've seen them too many times to remember. Leupp is a hot spot for UFOs." He went on to say that the UFO sightings at Leupp usually are of numerous UFOs traveling together in a triangle formation, but the sighting they had on January 24, 2007 was unique in that the craft flew alone and so low to the ground.

Deanna stated: "We see them, well, I don't want to say we see them all the time, but it's happed before. When they had that big sighting in Phoenix back in 1991, we saw the same lights here the night before. I guess it's just that no one ever calls it in or reports it here. This is pretty cool. I've never seen anything that detailed before, because it flew so close to us."

The local residence insisted that similar UFO sightings took place 15 and 17 years earlier in the very same location.

A similar sighting not far from Leupp was the Phoenix Lights of 1997.

The Phoenix Lights was a sighting back in March 13, 1997. The sighting took place over Arizona, Nevada, and the Mexico state of Sonora. The UFO sighting was even filmed by American Fox News TV station.

The most famous of all the eyewitnesses was Arizona Republican Governor Fife Symington. He says that he himself was a witness to one of the strange unidentified flying objects. He stated: "I witnessed a massive delta-shaped craft silently navigate over Squaw Peak."

At 6:55 PM on Thursday of March 13, 1997, an eyewitness in Henderson, Nevada was first to report a V-shaped craft with six large lights on its leading edge. It was coming towards his position from the northwest and passed overhead. In his written report to NUFORC (National UFO Reporting Center), he stated that it was the, "size of a Boeing 747." And then he went on to say that the UFO made an unusual sound like a "rushing wind." He noted that it continued on a straight path toward the southeast and vanished from his line of sight.

This unusual sighting would lead to massive eyewitnesses, numbering in the tens of thousands, and some estimate hundreds of thousands, over the next three hours of the UFO flying over Nevada, Arizona, New Mexico and Mexico.

A retired police officer also reported seeing the UFO over Paulden, Arizona. He had just left home and noticed it in the sky at 8:15 p.m. as it flew over a mountaintop. He was driving north when he gazed out the driver's side window looking west and saw a cluster of five reddish or orange lights. These lights consisted of four lights together, with a fifth light apparently trailing the other four. Each of the individual lights in the formation consisted of two separate sources of orange light.

At this point the officer instantly drove back home to get a pair of binoculars. He took them and watched the UFO as the lights disappeared over the horizon to the south. He said he watched them for a total of 2 minutes. He remembered that the massive craft made no sound at all.

At about this time, an overwhelming number of telephone reports started pouring into the NUFORC (National UFO

Reporting Center) and to other UFO organizations, to police stations, the news media and to Luke Air Force Base. These reports were called in from the communities of Chino Valley, Prescott, Prescott Valley, Dewey, Cordes Junction, Wickenburg, Cavecreek, and countless more from areas both north and west of Phoenix.

One group of three eyewitnesses just north of Phoenix said that they saw a huge, wedge-shaped craft with five lights on its ventral surface pass overhead with an eerie gliding type of flight. They watched as it traveled from the north and passed between two mountain peaks to the south. The eyewitnesses kept emphasizing one thing over all about the UFO, it was so massive that it was blocking out 70-90 degrees of the sky itself.

A second group of eyewitnesses to this UFO sighting was a mother and her four daughters near the intersection of Indian School Road and 7th Avenue. They were shocked when an object shaped like a sergeant's stripe came towards them from over Camelback Mountain to the north. They said that it came to a complete stop right above them and hovered over them for five minutes. They went on to say that when they looked up, the UFO blocked 30-40 degrees of the sky out and that it had a slight glow around its edges. They said that they could see individual features on the lower belly of the UFO and they were confident that they were looking at a single incredibly large, solid flying object.

The most important phenomenon that occurred that night was largely overlooked and ignored by the masses. The UFO continued moving south and then it suddenly fired a white beam of light at the ground. At the same time, seven lights on one side of the UFO shut off, disappearing from the eyewitnesses view. This is important because if the craft shot a white light at the ground, we can assume that it was not to destroy something, but rather to transport something or someone down to the ground, or up to the ship. What or who, we may never know, but can only imagine.

Another important report that night came from a young man who was an Airman in the US Air Force. He was stationed nearby at Luke Air Force Base, just west of Phoenix in Litchfield Park. He called in to the National UFO Reporting Center (NUFORC) at 3:20 on Friday. It was eight hours after the UFO sighting took place. He reported to NUFORC that the USAF F-15c fighters had been scrambled at Luke Air Force Base and had actually intercepted the UFOs. The USAF however would not confirm his story or the location of the fighter jets that night. The Airman gave highly detailed descriptions, which proved to be extremely accurate, based upon the information that investigators had reconstructed from eyewitness accounts. Two days latter, after his first call, the Airman called into NUFORC to report that he was just informed by his commander that he was being transferred to an assignment in Greenland. This Airman has never been heard of since that telephone call.

Perhaps they will never have the answers to what it really was flying over Leupp, Arizona on that evening in 2007, but one thing is for sure, these UFOs have been seen by tens of thousands of eyewitnesses, with video recorded and uploaded to youtube.com and other video sites. Although no major newspapers reported the Phoenix lights (clearly blocked out by the US Government) many small Arizona town newspapers did carry the photos and stories of eyewitnesses. Also Fox television News helped bring this sighting to the public, who bravery reported the UFO sighting on the spot in Phoenix, breaking the US Governments unwritten rule of not reporting UFO sightings. With bravery like Fox News and Arizona Republican Governor Fife Symington, we know as UFO researches, we are not alone in our endeavor to bring the truth to light.

Chapter Ten: 2007 UFO Sighting by Amateur Astronomer in Italy.

On April 29, 2007 Alberto Mayer who lives in Busto Arisizio, Italy and is an amateur astronomer found an incredible discovery almost by accident. He states that while adjusting the focus of his telescope that night at 10:52 pm, he spotted and tracked on video a UFO as it flew across the surface of the moon.

> 28. Photo of UFO recorded flying across moon. (At http://scwbook.blogspot.com/).

The short two-minute video he made seems more that of a professional astronomer than an amateur, but he prefers to be called amateur. The video shows the portions of the moon that he was focused on in close up mode while tracking a solid black round object as it flew over the surface of the moon at an elevation he measured to be 214 meters above the moons surface. This video is absolutely amazing, both in the way he put it together and in its organization. The round black item is tracked as he catches it on its journey beginning just after it passed Wallace crater and then passes over Wallace Alps, then on its way it passes below or near these craters, Egede, Eudoxus, Plana. As it passes Plana crater and comes between Daniell & Grace crater, it hovers and hesitates for about 15 seconds, before it makes a sudden change of path at an exact

45-degree angle. Also note that before it turned, 32 seconds into the video it abruptly slowed down to half its original speed before it suddenly hesitated for 15 seconds, and then moved upward at a 45-degree angle. As it moves upwards, you can see it pass the labeled craters on the moon in this order, Grace, Plana, Borg, Bally A, Bally, Gartner, Democritus, Moigno A, Baillaud, and De Sitter before it disappears around the moons edges to the dark side, out of the telescopes view. As the black UFO passes by Grace crater, it began to not only speed up, but also continue to pick up momentum until it was 1.75-2.0 times faster than it was when first seen traveling past Wallace Alps! This video can be seen on Youtube at http://www.youtube.com/watch?v=68J1kzZx65s by user name Exzero75, or you can see it at Alberto Mayer's official website at http://www.makina.it/SAA/Home.html . To view the video directly from him, you must got to the above website and then click on the word "movies" where you will be taken to several of his videos, one of which is the moon UFO footage in its highest quality offered.

Alberto Mayer stated he caught this UFO as, "I was focusing the scope while I noticed some dust spots on the camera CCD. I cleaned the sensor (twice) and after cleaning it, I noticed that new big black spot. I got really annoyed thinking that it was so difficult to clean that CCD then I saw the spot moving through the screen!"

At that moment he became excited and wondered at what it could possibly be. Quickly he had to act before it was lost in just minutes. He says, "Unfortunately the capture software was set to grab only 300 frame per sequence, so I had to restart the grabbing many times, losing some precious seconds between the movies. At the same time I must follow manually doing some mistakes while pressing the hand box slewing keys, due to the sudden excitement."

Alberto went on to say, "The Moon was 36° above the horizon not far to reach the Meridian. I thought it might be a geosynchronous satellite, but some calculations demonstrated that it should have been really huge! Now we are trying to investigate about the possible cause of this event. We guess it could be a balloon even if the size of the object seems not to fit with this option." Note this last sentence, because latter, another astronomer who only saw the video, insisted and persuaded Alberto to believe that it was in fact a balloon, which I will prove to you is impossible.

Here we find that Alberto is trying to find the answers and running calculations to insure he is correct. The subject of it being a balloon comes up, but is quickly dismissed because its size was out of proportion with the craft seen in the images. If it were a balloon, it would have to be floating at least at 6000 meters high. Note that passenger airliners travel at higher altitudes to avoid high winds and disturbances, this makes it even less likely that the UFO was a balloon due to the small chances that the astronomer would witness the object hesitate for 15 full seconds! Then change its direction at an exact 45-degree angle of what it was traveling earlier. This shows us it was not a balloon at all. The higher the altitude, the less turbulence and wind resistance, meaning a sudden change in direction seems too astronomically minute to even consider.

Imagine this, a weather balloon is sent off and it is like most weather balloons, made of highly flexible and thin latex material or even a chloroprene material might be used. Now lets place the luminous full moon on a clear night behind the thin latex or chloroprene weather balloon. With a telescope, what you will find is that it becomes translucent in its center and slightly darker around its edges. Such occurrences have been documented before and the translucent or clearness with the light shining through has been seen by other astronomers. Note, the UFO in the video that Alberto saw was a solid, flat black semi-round object. This object never demonstrated any translucency with the moon behind it, therefore it is reasonable to say Alberto was correct on his first assumption of it being an intelligently controlled flying object over the lunar surface.

Astronomers from around the world have reported such sightings as Alberto Mayer. They don't usually describe the object they see as a UFO as Alberto seems to have done in the beginning, but rather as an FMO (Fast Moving Object) moving across or near the lunar surface. Typically, these take the form of either light or dark spots, with velocities of 0*.001-0*.1/second and a duration of under one minute. This is very similar to Mayer's sighting where the duration was two minutes and the object was round and flat black. Scientists have evaluated the hypothesis that these phenomena are of terrestrial origin and only appear projected against the moon by chance, but statistical analysis reveals a different answer all together. It reveals to us that there are an overly amount of significant FMO sightings in the Mare Imbrium area near the Nectaris-Foecinditatis area. This is the same region that Alberto Mayer recorded the UFO.

This makes the April 29, 2007 by Alberto Mayer the best recorded, publically documented and distributed evidence of life existing on the moon to date. Although numerous buildings, structures have been seen on the moon, some even in the shape of faces, yet the governments around the world prefer to keep the public in the dark to prevent them from panicking. Even Alberto Mayer tries to tell others now, that what he saw on that day was just a balloon, because he knows if he ever wants to be taken seriously in the scientific community, he should not ever mention the words Unidentified Flying Object. His fear of being ridiculed by closed-minded individuals holds him back from declaring his true feelings and his true beliefs.

As I said before there have been numerous other sightings by astronomers themselves of objects that just defy explanation. Lets dive into just a few of the most memorable ones.

On February 2, 1990 in Hamburg, Germany an amateur astronomer observed a large triangle with luminous rings in its corners. Astronomer Gutschke saw the silent flying triangle when he was looking towards the sky at the stars. He calculated that the object was about seventy meters along each of its three edges. He noted that there was a luminous dust glowing in the night sky, even after the luminous UFO had already flown away. In each of the three corners of the UFO, were pink glowing rings. He recorded the altitude of the UFO at being 300 meters above the ground. In exactly eight seconds, the UFO had moved from zenith to horizon, towards the city of Hamburg, where it vanished.

In August 20, 1949 Astronomer Clyde Tombaugh had a UFO sighting. Understand that Clyde Tombaugh is not just another astronomer, but rather is the famous astronomer that discovered the planet Pluto. He said that he had seen several UFOs. He described them as six to eight rectangular glowing lights. These lights were window-like in appearance and yellowish-green in color, which moved from northwest to southeast over Las Cruces, New Mexico. He stated, "I doubt that the phenomenon was any terrestrial reflection, because... nothing of the kind has ever appeared before or since...I was so unprepared for such a strange sight, that I was really petrified with astonishment."

In 1956 Tombaugh said that he had often seen mysterious green fireballs that would suddenly appear over New Mexico in late 1948 with numerous sightings into the early 1950s. He stated, "I have seen three objects in the last seven years which defied any explanation of known phenomenon, such as Venus, atmospheric optic, meteors or planes. I am a highly skilled, professional astronomer. In addition I have seen three green fireballs which were unusual in behavior from normal green fireballs...I think that several reputable scientists are being unscientific in refusing to entertain the possibility of them being of extraterrestrial origin and nature."

Finally we come to my favorite UFO sighting by an astronomer. This UFO sighting was by Edmund Halley, the discoverer of Halley's Comet. In March of 1676 while looking at the stars he said that he saw a, "Vast body apparently bigger than the moon." In his recorded calculations, he estimated the altitude of the object to be 40 miles above him. He also wrote down that it made a noise, "Like the rattling of a great cart over stones." Using his genius skills with astronomical calculations, he quickly estimated the distance that the UFO traveled in a matter of minutes. He came to the conclusion that it had been moving at a speed of close to 9,600 miles per hour, which astounded him. Thirty-nine years latter in 1716 he would have a second UFO sighting where he witnessed it hovering in place for two full hours before it flew away.

In November of 2009, Vatican City, the same place that 400 years ago locked up the world famous astronomer Galileo for challenging the Catholic Churches view that the Earth was at the center of the universe, the Vatican has requested that experts in their field step up to study the possibility of extraterrestrial alien life and its implication for the Catholic Church. An astronomer and director of the Vatican Observatory stated, "The questions of life's origins and of whether life exists elsewhere in the universe are very suitable and deserve serious consideration." Rev. Jose Gabriel Funes presented the news stations with the results of a five-day conference that gathered astronomers, physicists, biologists and other experts to talk about the new up and coming field of astrobiology (scientific word for UFOlogy) which will focus on the study of the origin of life and its existence elsewhere in the cosmos. If the Church of Rome's views could shift so suddenly, then it leaves us to wonder, how long until scientists around the world give astrobiology (UFOlogy) the serious consideration that it clearly deserves? The future will not wait forever, nor will the truth.

Chapter Eleven: Astronaut Dr. Edgar Mitchell 2008 UFO Interview

Astronaut Edgar Mitchell is a brilliant scientist and an all-American hero who was born on September 17, 1930, (Note, same year as Apollo 20 William Rutledge astronaut). He was a pilot and an astronaut for NASA. Mitchell's crowning achievement that placed him into history was him being the pilot of the lunar module during the Apollo 14 moon mission. On this remarkable mission, he got a whopping nine hours of moon walking time along the Fra Mauro Formation. He holds the record for the world's longest moonwalk. On February 5, 1971, this moonwalk made him the sixth person to ever walk on the moon. A large number of photos began revealing unusual anomalies like photo AS14-68-9453 of Apollo 14, which showed a face carved in stone close up near an astronaut. Just Google the number and you will find it.

29. Photo of shadow of astronaut & big face on moon rock. (At http://scwbook.blogspot.com/).

In July of 2008 Mitchell did interviews on many news shows, TV stations and was on Kerrang Radio. On the radio show he said that not only was the Roswell UFO crash real, but that humans have been contacted repeatedly by aliens, but governments decided it best to hide the truth for over sixty years. He stated flatly that, "I happen to have been privileged enough to be in on the fact that we've been visited on this planet and the UFO phenomenon is real."

The radio station managed to get a statement from David Steitz, NASA headquarters spokesman, which denied the astronauts words saying, "NASA does not track UFOs. NASA is not involved in any sort of cover up about alien life on this planet or anywhere in the universe. Dr. Mitchell is a great American, but we do not share his opinions on this issue."

Furthermore, Dr. Mitchell goes to say that he absolutely believes in life on other planets and that he has no doubt about it whatsoever.

He stated to Kerrang Radio, "There is life throughout the universe. We are not alone in the universe at all." He said that he has been lucky enough to know firsthand that aliens have visited us on this planet (insinuating that he has had contact in person) not to mention that the UFO phenomenon is very real and has obviously been covered up by the US government for many decades. He said that there is more nonsense and rumors out in the real world than actual scientific evidence.

He insists flatly that aliens have visited Earth on several occasions and in his words, that the UFO events, "have been well covered up by all of our governments for the last 60 years or so," but some of the information has leaked out bit by bit to the public.

Although he was not born in Roswell, he grew up in and spent his early years in the area of Roswell, New Mexico, where Americas most famous UFO crash took place on July 7, 1947. He started school in Roswell and spent most of his youth and adolescents there before he went off to college. Because his family members were both farmers and ranchers, they knew the family that owned the land where the alien Roswell craft crashed. As a kid he only heard the stories from the older folks that were there at the time of the crash. He is confident that a real UFO had crashed because the military people and the intelligence people were very willing to talk to him once he had become an astronaut because they assumed that he was privileged to similar information. He believes that the remnants of the crash which included two aliens, one dead and one alive, were sent to some facility somewhere. But when asked where, he looked too scared to say. Instead he answered, "I don't know the answer to that one." He never mentioned Area-51 or Area S-4 just a few miles away from where he grew up. In an interview of unknown date with SciFi.com, Dr. Mitchell stated this about the Roswell crash, "the evidence points to the fact that Roswell was a real incident, and that indeed an alien craft did crash, and that material was recovered from that crash site."

If this is true, not only does the USAF have alien technology, but they also have the one live and one dead alien meaning they now have alien DNA to use for cloning purposes. Why clone a superior species? To raise it and gain its loyalty, and therefore use it for its extreme intellect as well as gain any knowledge engrained into its DNA.

Dr. Mitchell was fortunate enough to have been briefed on the UFO information periodically. He has been in military circles and intelligence circles of people from different countries whose job entailed things that go way beyond that of public knowledge. He has been involved with a lot of UFO work inside the government, but he says it's not his main work area. He has however been involved with research committees and research programs with very credible scientists and geniuses in their fields that do personally know the truth about aliens and UFOs that have been seen and even captured. Some of those people in the 1980s told him their stories of what they had witnessed in the government. He says that the government is currently in contact with aliens. He thinks a full disclosure of UFO information may come from the US government in the coming years. When he is asked by the radio interviewer if he was concerned about any possible severe consequences that the US government may use on him, he stated that, "no, I don't think they are knocking off anyone anymore. (Note, this sentence shows he was aware the US Government use to kill those that leaked UFO information.) Uh…doing drastic things to them." When questioned about the intent of the aliens being a hostile attitude or a non-hostile one, Dr. Mitchell felt that it was non-hostile for sure. He stated, "obviously if it were hostile, we would have been gone by now."

He says he has even been privileged to see aliens and says they are similar to the typical public belief in the fact that they have large eyes and large heads. He states, "little people that look strange to us." The aliens have small gray bodies. He says yes there have been ET visitations and crashed crafts and that there were materials and bodies recovered from those crashes. He also stated a few years back that, "there is a group of people somewhere, that may or may not be associated with government at this point, but certainly were at one time, that had this knowledge." He feels that the government's way of handling the UFO information leak is mostly based upon denial, by denying the truth of the documents, photo and films leaked and by trying to show that evidence leaked is a hoax or fake. He stated to SciFi.com that, "there has been a very large disinformation and misinformation effort around this whole area. And one must wonder, how better to hide something out in the open than just to say it isn't there. You're deceiving yourself if you think this is true, and yet, there it is right in front of you."

In his book, "The Way of The Explorer," published in 1996, Dr. Mitchell says that after he got back to Earth after the Apollo 14 mission that he was so mentally disturbed, that he could not remember the peek experience of his life. That he actually went to a therapist to get assistance to remember the things that he had seen through hypnotic regression and such. In his book he said that he conducted ESP experiments on the flight. (Understand that many advanced species communicate through ESP, like the Greys.) This plus the fact that after he finished the moon landing, he founded the Institute of Noetic Sciences, which primarily focused upon ESP, leads us to believe that Astronaut Edgar Mitchell, may possibly have been chosen for the mission because he himself has ESP or telepathy. This might be useful if they encountered the aliens know as the Greys, because they are said to be unable to speak orally, but instead speak telepathically. In his book he said that as he looked out toward Earth, surrounded by the blackness of space with an unfathomable number of stars, he felt, "an overwhelming sense of universal connectedness," marking the point that would change the direction of his life.

In the past, 1994, Dr. Mitchell had stated to the St. Petersburg Times Newspaper that there are many insiders within the US government that were studying the recovered bodies of aliens. That this specific agency has stopped briefing US presidents after John F. Kennedy.

Unknowingly when he said this, he added fuel to the fire about why Kennedy was really assassinated. One conspiracy theory that I discovered personally (SCW) is that Kennedy was assassinated because on November 12, 1963 he sent the CIA director a letter that ended with this statement, "When this data has been sorted out, I would like you to arrange a program of data sharing with NASA where unknowns are a factor. This will help NASA mission directors in their defensive responsibilities. I would like an interim report on the data review no later than February 1, 1964," signed "John F. Kennedy." It was a mere eleven days later when JFK was killed, making some question if the CIA resented taking orders from the president who they may see as, not having adequate experience to make such decisions. We may never learn the truth of who killed him, but we can assume that if no other presidents were advised of alien bodies, alien prisoners, alien captured or traded craft, then what happened with JFK that made them stop telling US presidents? Second, when JFK asked the CIA in the memo to share classified info with NASA, did the CIA believe that it was almost like going public with it?

A super secret branch of the CIA, privy only to a few, called the MKULTRA had a program of mind and behavior control research that they implemented. The program itself was begun in 1953 and would last until 1963. The CIA's first involvement with hypnosis came in the Office of Security, that in 1950 created special interrogation squads, each of which had an expert hypnotist. The hypnotist's reason for being there was to explore the subconscious of potential foreign agents and defectors from enemy countries. It went by the code name BLUEBIRD. The project was headed by Morse Allan, a former officer of both Naval Intelligence and the State Department. Allan asked young CIA secretaries to come visit him after their workday was finished. He quickly ran them through the hypnotic steps. He proved to his own satisfaction that he could control their every action. For example, he had secretaries steal classified files and give them to people who were total strangers. In this way the secretaries violated the most basic CIA security rules put out for them. He had them steal objects from each other and even started fires in the building. He had one of them go to the bedroom of an unknown man and then go into a deep sleep. Allen felt he had not gone to the edge yet, in 1954 he hypnotized another secretary, and told her to pick up a gun (unloaded) and shoot the other secretary, and she did. In 1963 the CIA released a 128-page "Counterintelligence Interrogation" manual. This document was not made public till 1997. Among its tactics it uses it sites hypnosis. This manual said that hypnosis "offers one advantage not inherent in other interrogation techniques or aids; the post-hypnotic suggestion." The document sheds a lot of light on the CIA and their planting of suggestions to forget, either what they had been through (like Dr. Mitchell entering an alien building or craft and communicating with them), or what they had done, or what they had discussed. The manual stated bluntly: "It should be possible to administer a silent drug to a resistant source, persuade him as the drug takes effect that he is slipping into a hypnotic trance, place him under actual hypnosis as consciousness is returning, shift his frame of reference so that his reasons for resistance become reasons for cooperation, interrogate him, and conclude the session by implanting the suggestion that when he emerges from the trance, he will not remember anything about what had happened."

Since the CIA has a branch that deals with UFO and alien technology and the fact that they don't tell NASA everything, indicates that the CIA may have interrogated Dr. Mitchell, either before the lift off or after his landing. A post-hypnotic suggestion could have been implanted for him to report to them everything after landing, where they then say the trigger word, causing him to forget everything that he had seen on his nine-hour moonwalk.

One thing is clear, nobody, especially someone as mentally and physically fit as an astronaut, would forget participating on a nine-hour moonwalk, not to mention that the astronauts air would not even come close to lasting for that long. The Apollo suite was only designed for seven hours of continuous use, yet it did have a 30-minute emergency oxygen tank. At http://history. nasa.gov/spacesuits.pdf a revealing discovery backs this up. On page 21 of the 28-page NASA document, it clearly states about the early Apollo spacesuits: "A backpack portable life support system provided oxygen for breathing, suit pressurization, and ventilation for moonwalks lasting up to 7 hours." The NASA made article shows Apollo 17 astronaut Schmitt in such a protective spacesuit while conducting lunar sample tests on the moons surface. The photo is dated 1972, one year after Dr. Mitchell's 9-hour moonwalk. If NASA's statement about the Apollo moon astronaut suits are correct, then we can conclude that for Dr. Mitchell to survive on the moon for that long, he must have spent 1.5-2 hours housed within a structure that had life supporting oxygen. The possibility that Dr. Mitchell might let such information slip out to non-government personnel may have been enough incentive for the CIA to enter the picture with their hypnosis techniques.

It is clear that around the world the personal testimonies of credible eyewitnesses regardless of their military rank or area of expertise, has been for all intents and purposes ignored. It is definite that there are tens of thousands of observable UFO sightings every year happening all around us, and those with the means to investigate are hording the information from the general public. Hero's like astronaut Edgar Mitchell, who put their life on the line to bring this information out into the open, are not given the publicity that they so rightly deserve. Did NASA have his experiences and memories that he experienced on the moon erased from his mind by CIA or NSA psychiatrists? We may never know, but what would it take for you to lose your memory of a nine-hour space walk and how would you feel afterwards if you had no memory of doing such an event? It may be that the things he saw on the moon caused him to suffer Posttraumatic Stress Disorder (PTSD). This disorder is pushed upon an individual during a traumatic event. A more complex form of the disorder is called C-PTSD, C standing for Complex. One of its main indications is significant mental amnesia and dissociates symptoms. If it was indeed C-PTSD that Astronaut Edgar Mitchell suffered, the things he saw or even touched on the moon during his walk, may have easily been that of an alien civilization. The nine hours of moonwalk alone are enough evidence to say he found something interesting enough to stay out there for quite a while, not the usual rock and dirt samples that one would expect, but perhaps instead, he was exploring actual ancient alien structures or ships, left behind from an advanced species eons ago. Still it remains, where was Mitchell during that nine-hour record setting moonwalk, and during the last hour and a half of having no oxygen? I am sure NASA's response would be, "No Comment."

Chapter Twelve: Hacker Enters NASA's Airbrush Room.

This particular sighting is even more authentic and more creditable since it came from the photographs of NASA's airbrushing room in Johnson Space Center. The room that airbrushes out any signs of life in the photographs including building, people, ships and machinery or anything that hints at intelligent life.

> 30. Photo of Johnson Space Center Sign. (At http://scwbook.blogspot.com/).

Although this persons sightings took place between the years 2001-2002, his continuing saga lives on as America presses the UK in June 17th, 2008 with its hardest hit manipulation of the political system yet. He not only saw UFOs on those computers, but also reports seeing alien technology for creating free energy, anti-gravity and found out the US has captured and back-engineered many types of alien spacecraft. He also seems to have stumbled upon an off world mission memorandum that the US Government has already begun. Gary is currently banned from using the Internet and faces up to 70 years in prison if extradited to the US. His overall goal was to bring information out into the open about the "UFO cover-up" and in many ways he succeeded.

Gary McKinnon is a systems administrator and is accused of hacking into no less than ninety-seven US military and NASA computers for two years. The cost of tracking down the hacker and correcting the problems that he allegedly created by breaking into their computer systems came to a whopping 700,000 US dollars, and another two million US dollars in fines, claims the US government. It seems strange that they say it cost so much to track him down, because the US government never says it took this amount of money to track down any suspect for any reason. (Makes the US sound desperate to get their grips on him). Also the US accuses Gary of causing damage, when he flatly denies causing any damage and says most likely they are referring to the cost of fixing the flaws in the computer programs that were suppose to keep him and other hackers out, but if was the governments fault for not having an adequate firewall or security program, then they should not charge him for fixing it. Often military pilots will point out flaws that need to be corrected in the stealth fighters and bombers, yet the government doesn't turn around and charge them with fixing those repairs. Gary denies causing any damage on any computers and says he took every measure of precaution to prevent such things from occurring.

Gary said he left a note on one of the government computers that went like this:

"US foreign policy is akin to government-sponsored terrorism these days…It was not a mistake that there was a huge security stand-down on September 11 last year…I am Solo. I will continue to disrupt at the highest levels."

Oddly enough, Solo was tracked down and arrested under the Computer Misuse Act by the UK National Hi-Tech Crime Unit. They informed Gary that he would defiantly have to do community service to pay for his crime, yet the Crown Prosecution Service in the UK refused to press charges against him. That's when the US government came in trying to get their hands on this hacker, not to punish him and throw him into prison, but to take him and blackmail Solo in order to force him to work for the USA. Such true stories have been made public on numerous accounts. The NSA or CIA will allow Gary to experience two weeks of actual prison like life, with numerous interrogations, and then when Gary is confident that he will spend the next 70 years behind American bars, they will present him with a choice. This choice will be to work off his time as an employee of the US government or stay in prison. They need people like Solo not only to hack into the military and government systems of other countries, but to prevent the tens of thousands of hacker attacks that the US system experiences every day. Gary McKinnon is too valuable of a resource to allow to go to waste and the CIA government is notorious for recruiting experts in their fields from around the world in order to keep America the leader of the free world. In 2009 CNN even wrote an article about how desperate the CIA was to recruit new hackers to assist them in their goals.

Gary was released and remained free until 2005, when the UK created a new extradition treaty with the US and an unnamed politically involved person in the treaty creation said that the entire treaty was created due to the US becoming desperate to get their hands on this hacker for his "controllable skills."

It's kind of sad really, because it's much like in the movie Star Wars when Darth Vader tells Luke Skywalker, "Come over to the dark side."

He goes by his hacker name of Solo, but his real name is Gary McKinnon. He is a British hacker that was being charged with "the biggest military hack of all time." Legal hearings in the UK have ruled in 2006 that Gary should be extradited to the US, but in February of 2007 his lawyers made and argument that he should be tried in England, not the US, but this was turned down immediately. On July 31, 2007 the House of Lords agreed to hear Gary's appeal and set the date for the hearing on June 17, 2008. The House of Lords delivered its judgment on Gary's fate on July 30th, 2008 saying that Gary can be extradited to the United States if they so wish. Gary appealed the ruling, but the appeal was rejected on August 28, 2008. If extradited to the United States, Solo faces a possible maximum prison sentence of up to seventy years.

Gary says that the movie, War Games was not his inspiration, but rather a book called The Hacker's Handbook, by Hugo Cornwall. He stated, "The first edition I read was too full of information…it had to be banned, and it was reissued without the sensitive stuff in it." Then he continued, "the book just kick-started me. Hacking for me was just a means to an end." Meaning that he was aware that governments were hiding loads of technology that could eliminate a lot of suffering around the world today if they were in use, but instead, they remain only in the hands of the US military.

Solo is being charged with hacking into the US Army, the Navy, the Air Force, The Department of Defense, and NASA among a few. Gary says that he did this because "I was in search of suppressed technology," or in other words, UFO technology. He said that it's the largest kept secret in the world due to it being linked to old sci-fi ideas and movies, making may people look upon such things as a joke, but that the technology they have hidden away is very important and could easily change the entire world over night. Gary is angry that elderly people around the world are struggling to pay their electric and gas bills and the US government is hiding technology that would allow the worlds population of elderly and children to have an endless and free form of energy to use in their daily lives. Gary soon found that the US government was sitting on suppressed technology for free energy and it sent him on his quest to find it in order to share it with the world.

He says that he broke into NASA and the Department of Defense easily because unlike what the press would lead us to believe, they are much less clever. "I searched for blank passwords, I wrote a tiny Pearl script that tied together other people's programs that search for blank passwords, so you could scan 65,000 machines in just over eight minutes." So he found computers that had a high-ranking status or administer status and haven't had their passwords set yet, but instead were on a password default setting. He says that there were lower than expected defenses to stop him from entering the computers and that he saw a permanent tenancy of foreign hackers beside himself on their networks. He could run a command when he was on a machine that showed connections from all over the world, and check their IP addresses from which they came from to see if they were other military bases checking in, but none of them were military.

Gary said he had hacked into the US computers over a period of several years and always went unnoticed because he was careful about the hours. He said he was always bouncing around many different time zones in order to stay hidden. By hacking computers during the late hours of night and earlier hours of morning, he insured that there would be less people around that could catch him.

He state, "There was actually one occasion when a network engineer saw me and actually questioned me and we actually talked to each other via WordPad, which was very, very strange. He asked me what I was doing? Which was a bit shocking. I told him I was from Military Computer Security, which he fully believed."

Before searching on the computer he researched for information on where he might find his UFO information, he soon found a group, and it went by the name of The Disclosure Project. Their website can be found at www.dislclosureproject.org/ and the group published a book that had 400 expert witnesses ranging from civilian air traffic controllers all the way to high ranking military officer and military radar operators, all the way to soldiers underground in the nuclear missile silos, whose jobs are to push that launch button if such an order came through. Gary said that they were very credible eyewitnesses that said they saw UFO technology, anti-gravity and free energy technology that have gone untapped by the public. Even scientist that have taken apart alien technology like ships and tried to find the reason of how it works. One of the most influence witnesses from the book Gary read was a NASA photographic expert. She said in her statement in the book that in building eight of the Johnson Space Center, they regularly airbrushed out images of UFOs from the high-resolution digital photographs taken from satellites. She said that there were folders called "filtered" and "unfiltered" and "processed" and "raw." This gave Gary the idea of a specific place of attack. The NASA airbrushing room!

Gary got onto his computer and easily hacked his way into that very airbrushing room at building eight of Johnson Space Center. He managed to get a picture out of the folder and what he saw on his computer screen was amazing. He stated, "It was a culmination of all my efforts. It was a picture of something that definitely wasn't man made. It was above the Earth's hemisphere. It kind of looked like a satellite. It was cigar-shaped and had geodesic domes above, below, to the left, the right and both ends of it, and although it was a low-resolution picture it was very close up. This thing was hanging in space, the Earth's hemisphere visible below it, and no rivets, no seems, none of the stuff associated with normal man-made manufacturing."

Gary believe that it was more than just coincidence that he found that photograph, since he read about the woman that said, "this is what happens, in this building and room of the space center." He said, "I went to that building, that space center, and saw exactly that."

He could not get a copy of the photo and print it out because his graphical remote viewer works only frame by frame and it was a Java application, so he could not save it onto the hard drive. Suddenly Gary was cut off as he was downloading the photograph. He stated, "I saw the guy's hand move across," (across the computer screen).

He also managed to get access to some Excel spreadsheets. He says, "One of them had the title, Non-Terrestrial Officers. It contained names and ranks of US Air Force personnel who are not registered anywhere else. It also contained information about ship-to-ship transfers, but I've never seen the names of these ships noted anywhere else." Can you imagine, US soldiers living on another planet, possibly permanently, to avoid them divulging the militaries programs and objectives to others on Earth? I the writer, have read on the internet about such a situation, it may or may not be legit, I could not confirm its source, but they say that there are over one thousand US Marines living off world and stationed in some alien built cities (possibly abandoned) on the dark side of the moon. Now, this seems unusual, but using the DVD camera technique that I mentioned in the moon chapter, I was able to find many cities, and even one entire crater that was covered in a dark sunglasses-like black dome. So, I for one am no longer laughing at such stories, but now looking deeper into them.

Gary understood what he was getting into when he started hacking the NASA computers. He says he knows, "unauthorized access is against the law and it is wrong." Yet, when he is asked what a respectable punishment for a person that has committed his crimes he responds, "Firstly, because of what I was looking for, I think I was morally correct. Even though I regret it now, I think the free energy technology should be publicly available. I want to stand trial in my own country, under the Computer Misuse Act, and I want evidence brought forward, or at least I want the Americans to have to provide evidence in order to extradite me, because I know there is no evidence of damage."

How important is free energy you ask? Excellent question and I believe US Air Force, Intelligence Operative, Master Sergeant Dan Morris said it best when he stated, "UFOs are both extra terrestrial and manmade. It's not that our government doesn't want us to know that there are people on other planets. What the people in power don't want us to know is that this free energy (from energy-generators developed with UFO technology) is available to everybody. So concealment of the UFOs is because of the energy issue. When this knowledge is found out by the people, they will demand that our government release this technology, and it will change the world."

Similarly, US Army, Colonel Thomas E. Bearden stated, "Probably 50 inventers have invented (virtually cost-free energy systems). If we use these systems, we can clean up this biosphere. But, what we have is a situation where the entire structure of science, industry, and the patent office are against you. It has been a victim of quite a bit of suppression. And behind this, we have a few people who are quite wealthy. The more powerful the agency, the more money they will resort not only to legal, but extra-legal means to suppress their competition. Lethal force is used."

If Colonel Bearden is correct, and had Gary McKinnon hacked into the patent office instead of hacking into NASA's airbrushing room, he may have been successful at finding the blueprints to create a device for free energy. Photographs of such machines would be much less valuable to the public than blueprints. One man, hacker though he may be, could have actually changed the world for the better. I wish they had a medal of honor for Joe Public when he literally puts his life on the line for not just his country, but the entire world. Gary McKinnon holds that great honor for fighting for those who believe in him and for those who do not. You are a modern day John Wayne. The world and I owe Solo a debt of thanks.

As of November 4, 2008, the BBC posted that the UK had 20 MP's sign an Early Day Motion that says that other countries can give sentence to anyone in their courts, but the person must serve out their sentence in their country of birth. This means Gary would be sent to America and tried in a court of law, but must return to the UK to server out any sentence he receives from the US courts. Justice Minister David Burrowes drew up the papers, saying that Gary suffers from a medical condition called Asperger's Syndrome. This is the law, but will the US government fight the UK further leaves to be seen. Solo would be a priceless addition to any government's arsenal of agents.

Update: On November 27, 2009, a decision to allow extradition was announced. In a letter sent to Mckinnon's lawyer, the UK Home Office stated, "The secretary of state is of the firm view that McKinnon's extradition would not be incompatible with his human rights," and that, "His extradition to the United States must proceed forthwith." After this, Gary's lawyer (Karen Todner) stated that she would seek a judicial review of the Home Secretary's decision. If this does not work, then they will make a new appeal to the European Court of Human Right's, because the court was unaware of Gary's Asperger syndrome when they first considered this case.

Chapter Thirteen: 2008 UFO over Turkey (Close-up View)

Over the past few years, Turkey has become a hotbed of sorts for UFO sightings. Hundreds of videos in 2008 alone revealed UFO sightings in cities such as Kumburgaz, Kurdistan, Istanbul, Kumburgaz, and much more.

31. Photo of Turkey UFO with two occupants in center. (At http://scwbook.blogspot.com/).

One such mass sighting of numerous UFO sightings took place in Turkey by an observant night guard working at a facility near the ocean. The sightings took place over several months and the videos the guard took are dated from June 22 to August 24, 2008. His close up videos were of what appeared to be several UFOs and the extreme close up of one of them that revealed not only the color of the metal, but surprisingly a clear window at its top, which was similar is shape to a stealth fighter, yet wider and was a single piece. Within sat not one but two slender figures. Dozens of eyewitnesses at the Yeni Kent Compound stated that over a period of a few months they had seen UFOs flying over the water along their beaches. The total of viewing time of his videos is suppose to take up to two and a half hours, but many fragments of it can be viewed across the internet in 2-9 minute segments at Youtube.com.

Yalcin Yalman witnessed the sightings several times before he had had enough and decided to bring his Canon video camera to not just record the distant dancing lights, but to also zoom in on the object to get a close up view of the irritating flying mystery. He was a night security guard in this area when he saw the lights. Everyday after that he brought his video camera in case he saw it again, and he was fortunate enough to get several more chances at getting some good footage. When he filmed them on his Canon camera with a Sony zoom lens, he had several people standing next to him that were also witnessing the unusual lights in the sky and conversing with him while he filmed. The camera that Yalcin used was a Canon DM-GR1-A with a 20x optic zoom that can go to 100x when maxed out. He added a Sony lens tele-objective 1758 model lens x1.7. I believe this camera to be at least three years old, because Yalcin said that it uses tapes, so it is assumed that it does not use DVD or stores video on a hard drive as the new 2008 video cameras currently do.

Yalcin Yalman does not speak English, but there are English subtitles of all his sightings, so that you can follow along easily. Each of the below sightings were caught on video by the night security guard. The videos show that each sighting was taken in PM but Yalcin said its because his camera and instructions were in English and he does not read or speak English, so he did not know what am or PM meant when fixing the camera's settings. All sightings he had, took place in the early hours of the morning along the beach. The sightings dates and times are as follows:

1. June 22, 2008   4:01 AM

2. July 30, 2008   4:50 AM

3. August 1, 2008   4:46 AM

4. August 2, 2008   3:59 AM

5. August 7, 2008   5:07-5:13 AM

6. August 10, 2008   4:20 AM

7. August 12, 2008   5:12 AM------Extremely close
                            view-2 ET's in cockpit seen.

8. August 24, 2008   1:44-1:45 AM

9. August 24, 2008   4:01 AM---Second sighting
    that morning.

Yalcin got such a close shot of the crafts that it literally filled up the screen of the video, showing its dull silver surface with a clear bluish window-like cockpit at its center. The UFO was estimated to be between 12 and 16 meters in diameter. The video shows the ship in the right angle as it slightly rotates, with the full moon behind it shining light through its clear cockpit canopy in its center. Also the craft does not appear saucer like, but instead looks more triangle shaped, but its corners are not sharp, rather the opposite. The corners are so rounded that at a distance the object has the look of being a disk till it rotates around slightly, in the video allowing the viewer to see that its shape is not morphing or changing, but has always been triangular in shape. Unlike a lot of triangle UFO sightings, this one had no lights flashing or otherwise. This video can be found at http://www.youtube.com/watch?v=oMlrMkHw55o sponsored by user Romulo282838 on October 20th, 2008.

On the video released are subtitles that viewers who only speak English can follow along with the conversation of the security guard and a few others watching the UFOs. As the security guard zooms in on the May 27th UFO he exclaims, "Oh my god, I got him again. It's moving very slowly. Moving slowly. It's right under the moon, near the sea." When the security guard looses sight of the UFO for a few seconds, he asks the person next to him where it went? They respond that it's over there, hovering over the sea. Using the conversation gives us some idea of what the others with him were feeling and seeing during this sighting. Although the sighting is extraordinary in video and seems to actually show not just one but several specific types of flying craft never before seen, the lack of other witness accounts leaves this sighting wanting. There were estimated to be over a dozen other witnesses of the UFOs flying over the sea, yet their accounts were never recorded and shared.

Kumburgaz is a very small city built alongside the Sea of Maramara. When looking out over the water, the only thing that can be seen is the water up to the horizon, yet there is a tiny island called Imrali Adasi far out of site of beaches of Kumburgaz.

Hmmm…let's examine this island a bit more closely using Google Earth Internet program. When we measure the island with Google ruler, we find that it is 6.31 km across and .72 km wide at its thinnest point, yet 2.5 to 3.02 km wide at its each of its ends. The satellite photo is semi-clear, but not clear enough to see the buildings very well. The island has only a few buildings on it and no boats are visible anywhere. Oddly, the buildings on the island look like no building that I have ever seen before. Instead they are silver tube-like in shape and smooth outside in places, built in a curvature in what appears to be a high hill. The tube gets thinner and disappears into the ground. Its size is 95.7 meters long before it turns and goes another 31.6 meters down, entering the ground. The building is 10 meters wide at its thinnest point and 31 meters wide at it widest point. Note that there are no airport runways visible and also there are no methods of transportation visible whatsoever. This very possibly may be a secret base used by the government to experiment with alien technology. It is very secluded, much in appearance of Area-51. This island using Google ruler puts at 49.39 km away from the city of Kumburgaz.

The different UFOs were recorded in the same location, by the same person every night over several months, so the

UFOs must have come from an area nearby, perhaps the island I mentioned earlier, but most likely if UFOs can travel in the depths of space, then most certainly they can also travel in the depths of the oceans. This leads us to believe that the small UFOs did not come every night from far away, just to hover over the Sea of Maramara for no reason. Instead we can conclude that they came from below the Sea of Marmara and the island itself is merely one of the entrances to this underwater facility. The idea of extraterrestrials living under our oceans is not a new concept. Stories of cities with advanced technologies have gone back thousands of years like the story of Atlantis itself.

The Sea of Marmara is only 142 miles by 44 miles wide, small in comparison with other seas.

1. The Kazakstan Government in Central Asia is said to be in the process of building the world's first alien embassy. This is according to a lot of leaked media coming out of Kazakstan. It is said that they have already as of April of 2009, allocated a large area of land in the city of Almaty for this project. Kazakhstan believes that open contact with aliens is imminent and by being the first nation to specifically create such a facilities, they are convinced they will reap enormous financial and economic rewards. They also see this as a great opportunity to demonstrate to the rest of the world, their forward thinking policies. The article states, "Currently it is generally accepted that aliens are making use of an underwater UFO base in the Caspian Sea, which Kazakhstan boarders." The head of neighboring Azebaijan's National Aerospace Agency, Fuad Gasimov has confirmed this statement to be true and he has stated on record, "that the old USSR constantly monitored alien spaceships regularly entering the water, but kept this a military secret."

Gasimov himself was part of this monitoring of the
UFOs for the USSR's science academy. Note, the Sea
of Marmara and the Caspian Sea are only 957 miles
(1,540 km) apart. It is said that ex-cosmonaut, Talgat
Musabayev (head of the Kazakhstan Space Agency)
is currently involved in the alien embassy project.
He is said to posses a large amount of information
on the aliens currently visiting earth. (For more
information on this alien embassy visit, Signs of
The Times, Kazakhstan Government to build UFO
base and alien embassy, by Michael Cohen. 4-9-
2009. http://www.sott.net/articles/show/181351-
Kazakhstan-Government-to-build-UFO-base-and-
alien-embassy .

Four months after taking the remarkable footage, Yalcin
Yalman showed it to Sirius UFO Space Science Research Center
to look into it further. The center studied the details for ten
days and came to the conclusion that it was genuine and not
a hoax. Sirius held a news conference at the Dedeman Hotel.
The sighting and video was shown to all who attended, and
was reported in both local and foreign media news of TV and
newspapers. The film was given to Professor Oktem so that he
could study it at the scientific institution owned by the state that
is highly reliable and influential, called "TUBITAK." This is the
science review board of Turkey. Professor Oktem took the film
to be analyzed at TUBITAK's National Observatory Center in
Antalya. Sirius and Yalcin Yalman were accompanied by TV
and newspaper press as they went and presented the film to
the professor.

The TUBITAK analysis came to the conclusion that the
flying object had a structure made up of a specific material and
was certainly not made up of any kind of computer animation
nor were there any forms of special effects used. They came to
the conclusion that the UFO sightings were most certainly real

and not a hoax. TUBITAK then stated, "It was concluded that these objects in the sighting that have physical and material structures, do not belong in any category (such as planes, helicopters, meteors, Venus, Mars, satellites, fire ball, Chinese lantern, ect...) but rather fall into the category of UFO."

The report then went on to talk about the video revealing a possible cockpit where passengers could sit in order to fly the craft. The report stated, "The light reflection from the left side of the object which is seen on August 10th shootings is not produced by the moon. At that time, the moon was in a phase that was pretty close to the new moon phase and located approximately at a 10 degrees proximity-angle to the horizon. Moreover, the image processing analysis conducted on some part of the footage revealed that the center of the object had the same density as its background, namely is of a transparent nature." Prof. Zeki EKER PhD, Director of the National Observatory of TUBITAK, signed this report.

In conclusion, I found the sighting to be real, although I question why a security guard has such a high-tech video camera topped off with an incredible Sony zoom lens on it. It's not exactly the hobby of choice of most security guards, so it leaves me wondering. Also the lack of eyewitnesses coming forward with their accounts of what they saw leaves this sighting hanging. There may be more eyewitness accounts, but they are probably still not translated into English as of yet. I found the islands nearby being possible alien/military alliance bases, but yet, any craft capable of flying through space should be capable of flying through water, making the island connected to an underwater base in this area the most likely conclusion.

Chapter Fourteen: US President Obama 2008, plus George W. Bush, Reagan, Carter UFO Sightings.

President Obama:

Lets start off with the current president. On November 1, 2008 the then President Elect (Obama) was giving a speech in Pueblo Colorado on his campaign trail to get elected as president, when MSNBC was filming video of Obama, they latter showed this video to the public, someone noticed an unusual object moving about 50-60 mph across the cloudy sky. The object appeared gray colored and without wings. There is no way that this video has been faked because MSNBC has the original, with the same UFO flying across it. How could someone at the MSNBC not notice this? Because, often UFOs are caught in digital photographs or digital video cameras, but the UFOs were not there when the person took the picture of the sky. This is because digital devices can see beyond the spectrum of our limited human eyes. Also note, the object is small and only takes up about 10 seconds of video, also making it difficult to catch. This means the crafts often have some kind of cloaking abilities, which would be in par with those who fly it since the pilots of such craft probably have thousands if not millions of years of technological innovations more than we do, yet again, the camera's lens in different than the human eye, and thusly can catch what the eye can not. This video can be found

at   http://www.youtube.com/watch?v=Hn8ZQ0xPgHY&feat
ure=PlayList&p=68335BFD6343533F&playnext=1&playnex
t_from=PL&index=36 and was posted by user PindzMedia on
June of 2009.

32. Photo of Obama & UFO in Colorado. (At http://
    scwbook.blogspot.com/).

Perhaps the above sighting was just coincidence, then the
chance of it happing again would very slim indeed. Wrong, it
happened again!

As nearly two million people gathered to watch Obama's
Inauguration (seen in the video http://www.youtube.com/
watch?v=-tLU07iNOm8&feature=PlayList&p=68335BFD63
43533F&index=37&playnext=2&playnext_from=PL), a UFO
sighting was made, and not by a spectator, but captured by a
cameraman who was broadcasting the scene live for CNN. Just
as the announcer was talking about how the Inauguration was
about to begin in thirty minutes or so, an object shot across the
Washington Monument as the camera zoomed out to show the
mass crowds that had gathered to see Obama. What was seen
is a grayish side view of what looks like a disk flying past the
Washington Monument, over the crowd and then vanishes
on the other side of the screen, far above the trees, which
discounted it being a possible bird. Also when looking at the
video in slow motion, I notice that people walking normal speed
will take only two steps before this UFO flies over them and out
of sight. This UFO crossed the distance of 110-130 meters in the
time it took the average person to walk two steps! The object
actually flies in front and past the Washington Monument and
over a CNN camera man hanging sitting mid-air on a swing
like apparatus hanging from scaffolding, making this object
about the length of a normal car from end to end. Now when
I freeze frame the video motion while the UFO is right over

the hanging cameraman, we see the UFOs length is equal the average automobile, even at such a great distance.

This video was later on FOX News and many other news agencies, each discussing this particular UFO sighting. Newscasters at FOX said "there was nothing in the airspace over the Inauguration that day. Airspace was closed."

UFOlogist around the globe called the sighting a 'rod.' It's a previously unknown life form that is active in our environment and is only seen flying.

Why you ask would an alien species be interested in the presidential inauguration so much that it would come to visit? One response overwhelms all others. At the inauguration were the most influential people in America today. People like Members of Martin Luther Kings family, Oprah Winfrey, Caroline Kennedy, Muhammad Ali, Al Gore, Jesse Jackson Jr., Beyonce, Jay-Z, John Cusack, Magic Johnson, Dustin Hoffman, Steven Spielberg, Samuel L. Jackson, Denzel Washington, Spike Lee, and Smokey Robinson. Most importantly also present were former Presidents Bill Clinton, Jimmy Carter, George Bush and George W. Bush, plus Obama. Now imagine all these people sitting together outside, without a roof over them, making them easy to view from a hovering craft and a temptation to aliens to come and investigate up close and personal.

The UFO could see a set of 5 US Presidents all at once and sitting together, outside and in full view of the world with the sky as their ceiling, sitting just feet from one another. If aliens wanted to see a mass of the most influential Americans, then this inauguration was the moment to do it.

Update to the above information. There has been a UFO sighting near President Obama. This time it was in 2009 in Norway on the day his plane landed there. Why was he there… to receive the Nobel Peace Prize. Why would a giant white and blue spiral be in the sky? It was aliens letting Obama know that they about the award he was about to receive and the spiral was to congratulate him. Read the 2009 Obama UFO chapter sighting in this book, "UFO Sightings of 2006-2009."

President George W. Bush's UFO Sighting:

President George W. Bush had his share of run-ins himself. On November 20, 2003 radar from the FAA and NORAD detected an unknown radar target within restricted airspace around the Whitehouse. The time was between 9:00-9:20 AM. Upon the UFO's discovery, preparations to evacuate the Whitehouse had begun. NORAD scrambled two fighter aircraft to the area that the UFO was detected, but nothing was found, and they called off the evacuation. If the US Government doesn't believe in UFO's then why are they running from them? Fox News, Reuters, and CNN covered the story.

This happened again on April 27, 2005. A UFO blipped across US security radar just 20 miles south of Reagan National Airport at 10:40 am. It moves swiftly towards the Whitehouse at speeds of about 75-120 miles per hour. The Whitehouse was fast to evacuate the president. Minutes latter President George W. Bush was in an underground bunker below the Whitehouse, not to mention that Vice President Cheney was not invited below ground (LOL), but instead had to be escorted off the Whitehouse grounds to a secure location. A Black Hawk helicopter was sent to the scene again, as well as a Park Police helicopter and a third local police helicopter for safe measures. All the helicopters that were sent to investigate, found only an empty sky. The object disappeared from radar only to reappear just a few miles from the airport again.

A lesser known sighting around President George W. Bush took place in Texas January 8, 2006. The people in Stephenville and Dublin Texas got a surprise appearance of a UFO in their area. Over a dozen people that included a pilot, county constable and many business owners state that they had seen a 300 meter long silent flying object that had brilliantly lit lights and flew low and fast to the ground. There's no need to tell you that this shook up the little communities terribly. Some of the eyewitnesses also said that they saw US Air Force jets in hot pursuit behind the UFO (Note, the jets may be escorting the UFO because the pilot may be President Bush himself). Radar reports that were released, show that eyewitnesses in Stephenville area were correct; something was seen on radar passing over them. The object was watched on radar as it flew over and past Stephenville and went directly toward President Bush's Crawford Ranch. At 10 miles from the ranch, the UFO went beyond the limits of the radar screen, or so the government agencies report. Leaving us with the unanswered question, how could any craft of such immense size come so close the Presidential ranch without getting a response from military aircraft? Unless of course, it was a military craft. MUFON Stephenville Radar Report, a 77-page document states, "This object was traveling to the southeast on a direct course towards the Crawford Ranch, also known as President Bush's western White House. The last time the object was seen on radar was at 8:00 pm. It was continuing on a direct path to the Crawford Ranch and was only 10 miles away."

Although humorous, George W. Bush was traveling to the West Bank, Israel, Kuwait, Bahrain, The United Arab Emirates, Saudi Arabia, and Egypt from January 8-16, 2008. While out there, he probably visited some secret US bases, making it highly likely he was out on a joy ride in some huge alien craft. America does have numerous bases over there, which could allow the USAF to experiment with alien ships without Americans noticing. So did the craft fly over the Crawford Ranch? That radar information was not given out to the public, but if it was headed in that direction on last radar report, most likely the answer is yes; it flew over the Presidential Ranch.

So if it was coincidence and Bush was not piloting, then were aliens perhaps looking for or working with the president? Or does a species from beyond, control American politics for their own gain? With President Bush being a former military fighter pilot, it may have been tempting for him to pilot such a craft as seen flying over his very ranch. Perhaps coincidence, or alien curiosity, only time will tell.

Think about this for a moment, former president George W. Bush decided to go skydiving on his 80th and his 85th birthdays. His 85th birthday was not his second, but actually his 7th time that he skydived! This guy is a risk taker. If the Bushes are that wild even in old age, then can you imagine what an ex-military jet pilot would do if he saw a 300 meter alien ship in the hands of the USAF? Remember that the US President also holds the title of Commander In Chief (Head of all the armed forces and has full authority outranking all other US military commanders). It is very possible that President Bush used his authority to take the helm of this ship from a hidden base and flew it half way around the world in mere minutes to take a look at his own ranch from the air. Please note, that he has already done this in the past, let me show you the proof.

On May 2, 2003 President Bush used his authority to co-pilot a Navy S-3B Viking hover jet and flew it over to the USS Abraham Lincoln aircraft carrier. Bush stated that he did fly the craft alone for a while stating, "Yes, I flew it. Yeah, of course, I liked it." His past experience of being a F-102 fighter pilot back in the Texas Air National Guard, had kicked in. It would be just natural for Bush to take control of an advanced alien ship in USAF hands and fly it to the location he loves the most, Texas. Why the ranch? Easy, to go home and impress his folks with his new ride. I'm sure former president George H.W. Bush was envious of his son and may even have got picked up at the ranch for a joy ride.

President Ronald Reagan's UFO sighting:

That brings us to Reagan's two UFO sightings. Reagan's first UFO sighting were made public by Steve Allen on his WNEW-

AM radio show in New York. Allen said that a well-known personality in the entertainment area had secretly revealed to him a UFO story that happened many years earlier. As the story had already been spread around the rumor mills, there was no doubt that the Steve Allan who was the host, was referring to Ronald Reagan and his wife Nancy. What was all the excitement about? Well, Ronald Reagan and Nancy were suppose to attend a casual dinner party with friends in Hollywood. All the guests had arrived except the Reagan's. They finally did arrive thirty minutes late and very disturbed by something. Ronald Reagan and Nancy nervously explained to the guest, "that they had seen a UFO coming down the coast."

Lucille Ball (Hollywood actor) who was at the event and overheard the Reagan's explanation and wrote a book called, "Lucy in the Afternoon." In the book she describes the event and says, "After he was elected president, I kept thinking about that event, and wondered if he still would have won if he told everyone that he saw a flying saucer."

The second UFO sighting that Ronald Reagan was eyewitness to happened in 1974, while he was still Governor. Just one week after witnessing the sighting, he told the story to Norman C. Miller who was the Washington Bureau chief for the Wall Street Journal, and later was the editor of the Los Angeles Times. Reagan stated to him: "I was in a plane last week when I looked out the window and saw this white light. It was zigzagging around. I went up to the pilot and said, 'Have you seen anything like that before?' He was shocked and said, 'Nope." And I said to him, 'Let's follow it!' We followed it for several minutes. It was a bright white light. We followed it to Bakersfield, and all of a sudden to our utter amazement, it went straight up into the heavens. When we got off the plane, I told Nancy all about it."

The pilot of Reagan's plane at the time was Bill Paynter, and he backed up Reagan's version of the UFO sighting. He stated: "I was the pilot of the plane when we saw the UFO. Also, on board were Governor Reagan and a couple of his security people. We were flying a Cessna Citation. It was maybe nine or ten o'clock at night. We were near Bakersfield when Governor Reagan and the others called my attention to a big light flying a bit behind the plane. It appeared to be several hundred yards away. It was a fairly steady light until it began to accelerate, then it appeared to elongate. The light took off. It went up at a 45-degree angle, at a high rate of speed. Everyone on the plane was surprised. Governor Reagan expressed amazement. I told the others I didn't know what it was. The UFO went from a normal cruise speed to a fantastic speed instantly. If you give an airplane power, it will accelerate, but not like a hotrod, and that is what this was like. We didn't file a report on the object because for a long time they considered you a nut if you saw a UFO."

Paynter said that it didn't end there. He said that he and Reagan talked about their UFO sighting every once in a while, years after the sighting occurred.

At President Ronald Reagan's speech that he read in front of the United Nations, he stated this: "Perhaps we need some outside universal threat, to make us recognize this common bound. I occasionally think about how quickly our differences worldwide would vanish, if we were facing an alien threat from outside this world. Any yet I ask you, is not an alien force already among us?"

President Jimmy Carter's UFO Sighting:

The sighting took place on January 6, 1969. Carter's sighting of the UFO began when the evening sky turned dark. The wind was still and Jimmy was standing outside the Lion's Club in Leary, Georgia. He was patiently waiting for a meeting to begin. From out of the blue, he and ten other eyewitnesses spotted a red and green orb pulsing in the western sky. Carter states, "it seemed to move towards us from a distance, stop, move partially away, return. Then depart. Bluish at first, then reddish, luminous, not solid." He said that sometimes, "it was as bright as the moon, and about as big as the moon. Maybe a bit smaller." He said that it began to move closer and closer to them. When the object reached the treetops, it hovered there for a while, then it disappeared shooting off into the distance. He said, "and none of us could imagine what it was, and I still don't know what it was." He believed the UFO was three hundred yards away from him at it's closest. The total time that they observed the flying object was ten minutes, before the object disappeared in the distance. After the sighting he was so taken aback by it that he decided to record his sighting orally on his tape recorder.

During an interview he gave with the Atlanta Constitution, Carter relayed how incredibly moving the event was to him. He called what he had seen that night a "very remarkable sight."

Jimmy Carter's mother Lillian admitted that her son had been in awe at the UFO sighting. She said, "the UFO made a huge impression on Jimmy. He told me about the sighting many times. He's always been a down-to-earth no-nonsense boy, and the sighting by him, as far as I am concerned, is as firm as money in the bank."

Carter has actually described the UFO sighting numerous times around the country since the day he witnessed it. In every single account, which includes the latest known telling of the story at Emory University, in the year 1997. Carter has always stood strong on his stance about the stunning nature of the UFO experience.

A few years after the sighting, at a Southern Governors Conference, he stated, "I don't laugh at people anymore when they say they've seen UFOs. I've seen one myself. It was the darndest thing I have ever seen. It was big, it was bright, it changed colors and it was about the size of the moon." In 1976, Jimmy Carter was quoted by the National Enquirer promising, "If I become president, I'll make every piece of information this country has about UFO sightings available to the public and scientists. I am convinced that UFOs exist because I have seen one."

Truly what Carter saw on that frigid clear night with 15 to 25 other witnesses from the Lions Club in 1969 must have left a great impression on him for him to come out into the public and readily admit to seeing a UFO.

Carter spoke to many government workers about the sighting. One was Press Secretary Jody Powell. When Powell was questioned about Carter's sighting, Powell said, "I do remember Jimmy saying that he did, in fact, see a strange light or object at night in the sky which did not appear to be a star or planet or anything that he could explain. If that's your definition of an Unidentified Flying Object, then I suppose that is correct. I would venture to say he has probably seen stranger and more unexplainable things than that just during his time in government." (Note, this last statement of Secretary Jody Powell's may have just unknowingly revealed that the US is involved with projects that have UFOs.)

In conclusion, we can clearly see with the evidence above, that those in control of the UFOs are highly interested in US Presidents, as well as future Presidents of the United States. Although President Obama did not see the three sightings himself that he was involved with, he was still the main attraction for them coming. Jimmy Carter was just one of many US presidents that said that he has personally witnessed seeing a UFO. Other presidents are Bush, Ronald Reagan (two sightings), President Eisenhower in 1952 working on a battleship saw a bluish type object and stood at the side of the ship watching it for ten minutes. Now imagine these people who have had UFO sightings being chastised for saying publicly what they have seen. It is ridiculous to even hold the presumption that a US president would lie about seeing a UFO. This is a person who leads a country that is looked upon as a world power. So why if presidents have seen UFOs, would there still be unbelievers out there? For the sake of humanity, do some research and keep an open mind and by all means if you ever see a UFO, please report it to the Mutual UFO Network (MUFON) or to some other on-line Internet UFO website and include a as much photographic, video evidence and statements of other eyewitnesses as possible. It is the documentation of the sighting that will help reveal the truth. Remember Jimmy Carter who went home after his sighting and recorded all the details on a tape recorder.

Chapter Fifteen: 2008 Indiana, USA.

On January 31, 2008 in Indiana a peculiar flying saucer was noticed in the evening sky. A witness who wishes to remain anonymous reported to MUFON (Mutual UFO Network, Mufon.com) that he was home alone and was feeling thirsty so he went to the kitchen to get himself a glass of water and maybe a bite to eat. As he got to the kitchen area he saw something that at first he believed to be a helicopter outside the window. It flew at about 300 to 400 feet away from his house. The object hovered just above the treetops across the street. The eyewitness lives in the countryside and thought it was highly unusual for a helicopter to be in his small community, and then he realized that he had not heard any noise coming from the object. He was surprised at this odd object and ran to grab his camera while almost tripping and falling over a rug in his walkway. As he got outside, he was amazed at the silent flying saucer that was above him.

33. 32. Photo of 2008 Indiana UFO. (At http://scwbook.blogspot.com/).

34. 33. Photo close up of China UFO. (At http://scwbook.blogspot.com/).

The witness stated, "I don't know how long it was there before I saw it, but it hung around for two or three more minutes." The

UFO was not moving or wavering in any way, but sitting solidly in the air. "The only thing I noticed was a sort of wavy-ness of the air surrounding the object. That's probably what stuck with me the most, actually. It resembled kind of what you see over a hot road on a hot summer day. It was getting dark, but I distinctly remember the dark tree line shimmering just below the object, against the sky-glow."

The eyewitness said that he had just enough time to take a single photograph before the batteries in his digital camera died. The person put down the camera and reached for his cell phone, capturing another photograph of the object. The UFO suddenly disappeared out of his cell phone camera view. The person looked up to see that the UFO moved directly away from the person at an exceptionally high speed and making no noise at all. It actually seemed to him to grow smaller, probably because of the sudden acceleration and incredible distance now between them. The UFO then curved upwards slightly going to a higher altitude, disappearing from view. The object flew in a direct westerly direction.

The witness stated that they had seen as a child some unusual looking lights in the night sky. It was from a very far distance and no detail except the color of the lights could be made out. The person had watched those lights do numerous odd stunts and tricks, making maneuvers that he believed no airplane could have possibly make. This was the only other sighting of a UFO that the eyewitness had ever experienced.

For more than two nights after the UFO sighting, the eyewitness was so shaken up about what they had seen that the person could not sleep. This in psychology is the mind trying to deal with what the eye had seen. The person is coming to grips with the terrifying fact that what he saw was real, yet seeing it in person literally blew him away, because UFOs are not suppose to exist except in sci-fi stories, or so the general public believes. Above all else, this statement clearly showed me that the UFO sighting that they had seen in Indiana on 2008 was a true sighting and not that of a hoax. The experience of seeing such an event, the kind we are taught since childhood that does not exist, can lead to temporary emotional shock. Thus leaving the individual feeling in awe at the object, but troubled about how it could exist, how governments tell us that UFO's are not real, yet there it is, hovering over your neighbor's home.

The 2008 Indiana, USA sighting was very similar to a past UFO sighting just two years earlier. In Nanjing, China on August 17, 2006 a similar (exact shape, color, lights and position of hovering over a building in residential area.) UFO was hovering nearby some apartment buildings. Note that in Asia, land is expensive in the big cities, much like New York, so apartment buildings differ in height, usually between 5 to 40 floors high.

A young Asian woman, walking out onto her patio, spotted the UFO. In the beginning of the short video clip, you can clearly see the apartments, which must be at least 12 to 20 floors high. A young woman begins to panic, calling to a man (Speaking in Chinese) to come and see what she is seeing. The woman sounds extremely worried and concerned in her tone of voice when speaking to the man about the flying object. The male voice clearly sounds a bit excited upon seeing the UFO. The UFO has no sound in the video so we can assume it is relatively silent both while hovering and when it suddenly accelerates at incredible speeds from a standing position.

The video footage that the young woman took only takes thirty seconds of video before the gray saucer suddenly powers up. Seven small glowing white lights suddenly turn on. (Note, the Indiana UFO had same lights, color & spacing of lights.) The lights don't only turn on, but if you look frame-by-frame in slow motion, you will see that the lights grow in intensity, with a fuzzy waviness' appearing to grow on the left side of the ship. This waviness' grows as the seven small lights suddenly burst so brightly that the gray UFO turns into a white cloudlike image and disappears moving at incomprehensible speeds to its left.

The woman and the man in the video are speaking Chinese and the interpretation is as follows.

> Woman: "Look, look, what is it? What is it in the sky?
>
> Man: "Over where?"
>
> Woman: "Over there! Over there! What can it be?"
>
> Man: "Hm."
>
> Woman: "Over there? Is that a UFO?"

1. This video can be found at, Youtube.com. UFO in Nanjing, China. By user 'qxynodoubt.' http://www.youtube.com/watch?v=DbSSTqlanS8

2. The 2008 Indiana UFO at MUFON. Photo of 1-31-2008. http://www.mufoncms.com/files/9511_submitter_file1_IMG_4232resizedForNet.jpg

The UFO in Indiana and the UFO in Nanjing, China is clearly the same type of craft as can be seen in the photographs when compared side-by-side. We can assume when looking at the photograph of the Indiana object, that it was at the time of the sighting, ready and powered up for shooting out of there at high speeds. The only difference is that in Indiana the craft hovered for several minutes before shooting off. It is apparent that the seven lights on the sides of the craft are not really lights, as on an airplane, but instead are some kind of generators for propelling the ship at high speeds. The question about the Indiana sighting that bothers me a bit is, why did it hover for 2-3 minutes while powered up? Perhaps it was merely pilot's hesitation to give or initiate the take off. A simple distraction in the ships cockpit could easily distract its pilot for a few moments. The ships diameter is difficult to calculate because of its distance away from the building (China sighting), but appears to come close to thirty-two yards across or 106 feet. (Note, 108 feet UFO saucer was photographed under hanger in Area S4 chapter.) A ship this size could possibly hold a crew of up to twenty. So what was the craft doing in Nanjing, China in 2006 and Indiana, USA in 2008 flying over public neighborhoods? Whoever the pilot was, he must be very curious about the average humans and how and where they live.

Chapter Sixteen: Texas 300 meter long UFO, 2008

Try to imagine if you will, you are walking though a field enjoying a beautiful clear day, when all of a sudden a metallic gray cigar-shaped craft that measures close to 300 meters in diameter slowly blocks out the sun above you and flies over you at an altitude of only 100 meters, and to top it all off, afterwards the US Air Force calls you delusional, saying what you saw was just reflections of lights bouncing off aircraft and clouds. Pareidolia is a bluffing technique the USAF depends upon. Pareidolia is when a person notices an illusion or misperception involving a vague and obscure stimulus such as an image or sound as being significant. Much as anyone may look into the clouds and see shapes like animals, faces and such.

This of course would make you feel more than a little peeved at the USAF, but try to understand that America has an agenda (unlike some other countries like the UK) to hide such sightings for military reasons beyond us, or so they believe. You would think that if over a dozen people saw a UFO, then there would be more believers. Not so with the Texas sighting. I noticed a lot of effort was being made to conceal this story and downplay it as much as possible. As I told you earlier, when the debunkers conglomerate on a story, then there is serious cause to believe it's real. It's also interesting that the US Air Force made a statement that clearly called the twelve eyewitnesses liars, yet a month latter the US Air Force took that back and admitted to having F-16 fighters on a training mission in the area that the witnesses were in. Media frenzy soon formed and gave this sighting worldwide representation, which is needed since the American government tries to downplay most sightings as illusions or airplane lights. Silly rabbit, tricks are for kids.

On January 8, 2008 the people in Stephenville and Dublin, Texas got an unforgettable eye opening experience in their area. Over a dozen people that included a pilot, county constable and many business owners state that they have seen a large silent flying object that had brilliantly lit lights and flew low and fast to the ground. There's no need to tell you that this shook up the little communities terribly. Some of the eyewitnesses also said that they saw US Air Force jets in hot pursuit behind the UFO. This strange cigar shaped craft was certainly too large, quiet and fast to be an airplane. Many said that the giant craft had lights that frequently changed configuration. The locals reported the flying craft for several weeks in the towns of Stephenville and Dublin. (Note: Apollo 20 chapter and the alien cigar shaped ship they wanted to retrieve in Deporte Crater. Perhaps this is it after it has been repaired.)

Weather conditions at the time of the sighting on January 8th, 2008, were clear and the sky had no clouds, allowing people to have a clear view for more than ten miles. The temperature was in the 40's in Fahrenheit, with a slight wind. Sunset fell at 5:44 PM.

A former air traffic controller was among the list of eyewitnesses to the UFO. He was in the west of downtown Comanche. He had a description that was similar to that of Constable Lee Roy Gaitan. He had seen many lights that were moving around in no particular pattern. It lasted for one minute, and then they disappeared altogether, like turning off a switch. About twelve minutes latter he saw many military jets in the same area where the UFO had been, and he compared them in size to the UFO as being like raisins to a grapefruit. Other eyewitnesses said it was 38 times greater than that of the fighter jets. Since an F-16 is 49 feet long, we can conclude that the UFO was about 1862 feet in diameter or 562.4 meters.

One eyewitness was Ricky Sorrells who was standing in a field with many leafless trees around when suddenly the light above was blocked out by a "tin barn gray" metal emitting unusual mirage-like heat waves (antigravity waves, refer Area S4 chapter) emitting from large holes on the bottom of the craft. These holes were crater like and were evenly spread throughout the bottom of the craft at distances of 40 feet from one another. Ricky stood breathlessly watching through the trees as this mysterious object float silently above him. He believed the crater like holes in the bottom of the craft were about six to eight feet across with the inner part of the crater appeared to be closer to three feet in diameter.

It is the writer's personal belief (SCW) that these craters on the bottom of the craft were actually antigravity generators that created the heat like waves that Ricky saw that day. Remember that this is an enormous craft that may be considered a small mother ship, so it has a lot of area of the 300 meters in diameter to use antigravity generators to give it flight. Remember in the Area S4 chapter that these antigravity generators use element 115 and that the elements size is smaller that a half dollar yet one of these coin size pieces would power three antigravity generators for close to 20-30 years of use.

Notice if you will that larger flying sauces seem to only be seen in dense populated areas like the desert and over the ocean rather than over New York City or Los Angeles, which leads me to wonder if the US Air Force has a agreement with the aliens themselves to only fly in designated areas of Earth.

The one thing that was missing from this sighting was video or photographic evidence to confirm it. It didn't take too much time before someone in Texas to reveal a video of the UFO. Virgil Fowler sent video footage of a UFO to the Austin, Texas, FOX television station. Soon afterwards the video spread across the Internet and television news. Although the video of the giant craft was taken over Lake Travis in Austin, an entire region of Texas has been reporting UFOs, most reporting the same 300-meter craft that those in Stephenville and Dublin had seen, but check out this video news cast of the UFO in Texas, January 2008. By user 'DesignJZ.' http://www.youtube.com/watch?v=kyECwMsJdbE .

The story first made the headlines when Peter Davenport of the National UFO Reporting Center was a guest on the Coast-to-Coast am radio show, and he released several reports from the Stephenville area. Steve Allen, a veteran pilot, made the first report and he talked about his experience on ABC News saying:

"Flashing strobes, silent, flames out the back side and jets chasing it! On 01-08-2008, at 6:15 PM, CST, my friends and I were watching the sunset when several strobes or flashing lights coming from the east at about 3500 feet and heading west toward Stephenville, Texas were observed. Estimated speed was 2000 to 3000 MPH."

Since an F-16 can only travel up to Mach 2 or 1,500 mph, it should have made for a fun chase for the military. None of the F-16s fired any shots at the UFO, but this seems a typical US military strategy in most F-16, UFO encounters, perhaps indicating a don't fire unless fired upon strategy.

Steve Allen went on to say, "The strobes made several changes of flash patterns and configurations. The flight duration lasted about 3 minutes. The front two strobes were about ½ mile apart and the back ones were about 1 mile back from the front of the strobes. The backside of the flashing lights came to a vertical flashing. Then there were 2 separate vertical flames about ¼ mile apart for several seconds and the craft was gone. We never heard any noise from the craft! They headed west towards Abilene, Texas. Then about 10 minutes later, the craft was seen again. This time with 2 jets chasing it. They were headed east toward south Ft. Worth, at about 4000 feet altitude.

The jets were unable to catch up with the UFO and went off into the distance at full throttle. I am a pilot and have been flying for 30 years and have never seen anything like this in the area. The craft is not from around these parts!"

Multiple eyewitnesses confirmed that they also saw the two military jets in hot pursuit of the alien craft, which soon pressed the military into admitting more, even if it wasn't everything. The Air Force Reserve admitted that on the night of January 8, they had ten F-16 fighter jets that were holding training flights in the Stephenville area that night. Originally the 301st Fighter Wing at the Naval Air Station Fort Worth Joint Reserve Base stated, "none of our jets were in the area that night." The Wing spokesman latter reversed their words and said that, "yes they did make a mistake and jets were flying in the area where eyewitnesses attest them to have been."

One of the most detailed reports from a witness about the UFO buzzing around the area was Constable Lee Roy Gaitan. Lee Roy lives with his wife, son and daughter just south of Dublin. Lee Roy has been Constable for four years and a police officer for seventeen years. He not only investigated this sighting, but he and his son were eyewitnesses themselves to the strange UFO that flew over them on January 8, 2008. On that Tuesday at close to 7 PM, he had strolled out toward his car to get a credit card so he could rent a Direct TV movie for his family. The forty-three year old Dublin law enforcement Constable said that in all his life, he had never seen such strange flashing and colors of lights in the sky. He said, "As I'm walking back to the house, I noticed first a red glow in the southwest sky. It was like a fiery red glow. It wasn't really like a light. It was more like a fiery-looking ball." He continued saying that the craft was about 300 feet up, based on the tree line and it was close to 500 yards from him. It was not moving. It appeared to be hovering. As he stared at it, it slowly faded away before his eyes (possibly turning on a cloak). To his amazement it reappeared again, almost in a horizontal line from where it had once been, but now was a little further to the west. Lee Roy saw it fade as if you were slowly turning on and off a rheostat switch. But it didn't look like a light, but instead it was a color like Lee had never seen before, a fiery red and not a bright red. Around it was a yellowish glow. He then went back inside his house and told his wife about it. She gave him an odd look. Lee Roy quickly goes back outside followed by his eight-year-old boy and they look up into the sky, hoping to see it again. Then at about 4000 feet altitude they see white flashes of light. These were extremely bright like LED lights and not like a flash, but more like a strobe. These lights were not staying stationary in the sky, but instead were bouncing around to different locations. They seemed to follow a pattern where one light would go on and then off, then another one would come on and go off. He went over to his car and took out a set of binoculars to try to get a better view of the craft. It was hard because the craft kept jumping about, but finally when he did focus on one, he could see nothing but light. There was no outline, nothing seemed to be attached to the nine lights that were spread a good distance apart from each other. Suddenly as if all the lights were attached to the same string, all at once, they rocketed off to the northeast sky at a speed so fast that it seemed impossible to him.

Note, this may indicate that the lights appear as separate UFOs, because the ship itself may by incredibly large and invisible to the naked eye, yet its maneuvering thrusters may be what he was seeing. What I am saying is, Roy's saw lights flashing on in one place then off and appearing in a different place. I am suggesting they are all connected to the same ship, and that is why at the end when they all shot off, they shot off together as if they were one. There is technology today that allows light to bend around it to make things appear invisible. Check out this demonstration video, Optical Camouflage (Invisible Cloak) at http://www.youtube.com/watch?v=JKPVQal851U by user RakanGG. This video demonstrates new invisibility technology.

Roy stated, "I had trouble keeping up panning my binoculars. It was that quick and they were out of sight. I'm sitting there thinking, what was this thing?" He was stunned at its speed, thinking that no human body could withstand the G-Force that these things used when they shot off.

Lee Roy said that on Saturday, January 12, 2008 he got to see a video that was taken from the in-the-dash camera of a car. The video had footage of the UFO that the officer watched that day. When Lee Roy viewed the video he saw the alien craft changed forms three different times. The first time, it was a white glow moving very slowly. Then it changed from white to green to blue and red lights. It looked almost like a strobe light.

Lee reported that a police officer on patrol had radioed in to inform him that he had seen a huge cigar shaped aerial craft with two tall antennas topped by red lights moving slowly about 300 to 400 feet in the air above Stephenville, Texas courthouse. The police officer said it was about three football fields in length. A computer rendition was made of the object and it confirms the shape of the craft seen in the video mentioned earlier. The police officer with the video said, "keep watching," so he continued to watch for another eight minutes. Then he saw something unexpected, the UFO changed shape to look like a jellyfish. (Note: Jellyfish UFOs have been seen over Russia multiple times.) He said it may even be compared to the shape of a parachute and it was very bright white. After two minutes more the craft changes vertically, like a line going up and down. The line appeared to be solid white no longer flashing. It stayed in that shape for a few more minutes. This thing was moving and the officer was having trouble focusing the camera. Then the police officer told me he and another officer saw another craft just like this one about a mile behind the ones he was watching with the camera. This camera zooms in close to 600X or six times larger, but you still can't see any type of craft. In all, Lee Roy watched the tape for 16 minutes, before the craft shot out of sight.

He says that almost a half dozen police officers had seen the UFOs over a few day period, but that they were too afraid to come forward for fear of losing there jobs or being ridiculed. He says he was not afraid of his boss because in his position as Dublin Constable, he doesn't have a supervisor over him, but instead the good citizens of Erath County are his bosses. So it would take thousands of people for him to lose his position.

On Larry King Live, Stanton Friedman, world-renowned UFO researcher and Physicist said that what people saw in Texas was too large to be an average UFO, but was instead a mother ship or space carrier. When Larry King asks Friedman about the military jets chasing the UFO craft Friedman stated, "Well, that's standard practice. Larry, I hate to say this and it sounds unpatriotic, but the Air Force has lied about UFOs for 50, more than 50 for 60 years, actually. And this would be in keeping with that." Then he went on to say. "They want to figure out how they work. They worry about somebody else figuring out how they work. If they were to announce some UFOs are alien spacecraft, I would expect there would be a push toward an Earthling orientation, as opposed to American, Chinese, whatever. I don't think any president wants that." Latter he said that a lot of people ask him why the aliens don't just come down and land on the White House lawn, and he always tells them that the president of the United States, does not speak for six billion Earthlings. Often he doesn't even speak for 300 million Americans.

MUFON representative in Dallas, Texas Ken Cherry believes this sighting to be the far most significant mass sighting since 1997 in Arizona when hundreds of witnesses saw what is today called the Phoenix Lights.

In the first two weeks of February of 2008, there have already been reports of two more disc-shaped aerial lights and spherical plasmas at the infrared deer feeder cameras in Brownwood, Texas. (Note that in 2007 an unusual photo of an orange glow was taken opposite a deer when a wildlife camera took its photo.

In that photo, the glowing orange orb had two eye sockets, one nasal cavity and a mouth. In total the expression on the glowing orbs face appeared surprised at the sudden flash bulb of the wildlife camera.) Lee Roy investigates into these matters since he is the local law enforcement.

Radar reports that were released show that eyewitnesses in Stephenville area were correct; something was seen on radar passing over them. The object was watched on radar as it flew over and past Stephenville, Texas. It flew directly towards President Bush's Crawford Ranch. Radar kept sight of it till it was ten miles the Bush ranch. Then radar reports say…the object left the radar screen, no longer being able to be tracked. Leaving us with the unanswered question, how could any craft of such immense size come so close the US Presidential ranch without getting a response from military aircraft? Unless of course…it was a military craft.

MUFON Stephenville Radar Report, a 77-page document states, "This object was traveling to the southeast on a direct course towards the Crawford Ranch, also known as President Bush's western White House. The last time the object was seen on radar at 8:00pm, it was continuing on a direct path to the Crawford Ranch and was only 10 miles away."

Although humorous, George W. Bush was traveling to the West Bank, Israel, Kuwait, Bahrain (possible secret US base here), The United Arab Emirates, Saudi Arabia, and Egypt from January 8-16, 2008, making it highly unlikely he was out on a joy ride in some huge alien craft. So did the craft fly over the Crawford Ranch? That radar information was not given out to the public, but if it was headed in that direction on last radar report, most likely the answer is yes; it flew over the Presidential Ranch. So if it was not a coincidence, were they perhaps looking for or working with the president? Or does a species from beyond, control American politics for their own gain? With President Bush being a former military fighter pilot, if was told about the US Air Force having an alien mother ship, it must have been extremely tempting for him to pilot such a craft, such as the one seen flying over his very ranch.

Let me restate what I said in an earlier chapter about Bush. Former president George W. Bush decided to go skydiving on his 80th and his 85th birthdays. His 85th birthday was not his second, but actually his 7th time that he skydived! This guy is a risk taker. If the Bushes are that wild even in old age, then can you imagine what an ex-military jet pilot would do if he saw a 300 meter alien ship in the hands of the USAF? It is very possible that President Bush used his authority to take the helm of this ship from a hidden base and flown it half way around the world in mere minutes to take a look at his own ranch from the air. Please note, that he has already done this in the past as proof. On May 2, 2003 President Bush used his authority to co-pilot a Navy S-3B Viking hover jet and flew it over to the USS Abraham Lincoln aircraft carrier. Bush stated that he did fly the craft alone for a while stating, "Yes, I flew it. Yeah, of course, I liked it." His past experience of being an F-102 fighter pilot back in the Texas Air National Guard was kicking in. It would just be natural for Bush to take control of an alien technology ship in USAF hands and fly it to the location he loves the most, Texas. Why the ranch, easy, to go home and impress his folks, namely George H.W. Bush with his new ride.

It's clear that the sighting in Stephenville and Dublin, Texas were by competent individuals that clearly saw unidentified craft of enormous proportions in the sky. Lee Roy said that the Air Force didn't send anyone to investigate the video footage that he had in his possession, which tells me that either the US Air Force got closer footage from their F-16 fighter jets taking video, or the craft is owned by the US Air Force, meaning since such craft seems impossible to make today, that they most defiantly have such a craft from an alien treaty or agreement, something that most governments work towards, in hopes of gaining military superiority. It's easy to dismiss something like this as guessing, but understand, if aliens do exist, what do you think the United States government would do to get their hands on their technology? Cultures thousands if not millions of years older than our own would have a lot to offer our culture. I personally feel that it's not a military craft, but an alien craft buzzing the area for future mission sites, under what agenda, only they would know.

On January 11, 2008, just three days after the Stephenville UFO sighting, another sighting of a much smaller craft took place just sixty miles away in Denton, Texas. MUFON (Mutual UFO Network) reported that an eyewitness saw the craft flying along the shores of Lake Ray Roberts. The person stated, "It was flying very slow, approximately 30 mph and 30-50 yards away. It seemed to be hugging the shoreline, flying at eye level. It made no sound at all." The nameless eyewitness said that the craft was black and gray in appearance and was triangular or delta-shaped. On his way to work at 6:50 AM, he drove along bear road where he first saw a bright white light-flying coming over a hill near the dam. The eyewitness had been in the Navy and had had experience with military aircraft, but this was a craft that he had never seen before. He slowed down the car and opened the window to get a better look at it. It was no bigger than his car. Its rear was rectangular, with three round openings within it. He could not see any flames or exhaust come from it. The top of the UFO had a canopy shape that was long and thin. The canopy appeared metallic. The bottom had panel lines or seams. It had three white lights, one in each of its corners, glowing constantly without blinking. (Note, this ship has seams and is in the shape and color of a stealth fighter. This is not an alien craft, but a USAF new version of its stealth fighter with anti gravity engines fitted into it. This increases the possibility that the mother ship seen on January 8, 2008 in Stephenville may be controlled by the USAF.)

On January 8, 2008 the Denton Radar data showed a total of ten military fighter jets from CAFB (Carswell Air Force Base). On radar the jets flew in three separate groups. Two groups of four and a group of two. Radar reports also show an AWACS (Airborne Warning and Command System) aircraft in the area of the UFO sighting. This AWACS aircraft was in the area for over four hours, while the fighter jets were only there for a total of seventy minutes. (Note: This AWACS plane my indicate they USAF knew the UFO was coming.) Using the radar information, the UFOs speed was seen to hit 2100 mph. The UFOs speed reached 532 mph in 30 seconds, then ten seconds latter it suddenly slowed to 49 mph.

This sighting in Denton, Texas is too close to the Stephenville sighting not to be connected. The strange thing that stands out about this MUFON report is that the unknown eyewitness says that the craft had visible seems in its bottom. That means the craft was made of smaller panels to make the crafts bottom. This does not conform to millions of UFO close-up reports saying they were seamless, top and bottom. Therefore it can be concluded that the Denton city UFO was merely one of the USAF experimental crafts designed in Area S4, and conforms to the next generation of stealth fighter shapes that they currently have already in use. Although designed in Area S4, the craft was over 750 miles away, meaning its home military base could be…anywhere.

Chapter Seventeen: 2008 Three Human-Like Figures on Mars, One Giant Reptilian Head With Crown & Three Carved Faces In Side of Mountain.

On January 3, 2008, NASA posted a photograph taken by the Mars Spirit Rover on their web site that bloggers around the world began scanning over carefully in order to catch anything that NASA may have accidentally let slip through. To everyone's amazement some anonymous person in Asia did and posted it on the Internet causing widespread excitement over the possibilities of life existing on Mars. This led me to new discoveries of a male and a possible child figure near the female figure. I also found a statue like face (head only), in high detail of a reptilian looking skyward and wearing a crown with a jewel in it (I named it Waring 1). I found a three carved faces in the side of the mountain of three unknown species, which can only be compared to Mt. Rushmore (I named it Waring 2 for the record).

35. Photo of three figures on Mars. (At http://scwbook. blogspot.com/).

36. Photo of three heads on Mars. (Waring 2) (At http:// scwbook.blogspot.com/).

37. Photo of large Reptilian head with crown on Mars. (Waring 1) (At http://scwbook.blogspot.com/).

When I found the NASA photograph at http://photojournal. jpl.nasa.gov/jpeg/PIA10216.jpg , I was amazed at the quality. The figures themselves were in the far left side center of the photo. I realized NASA labeled it 'False Color' so I decided to change the color back to its original state to see what NASA was hiding. When I used auto balance for color adjustment from Photoshop, the colors changed slightly. The color of the tunic like cloths turned from brown to a dark olive green, the faces from brown to pinkish. I say faces because although news reports only one figure in the photo, there are actually two or possibly a third (that of a child). That's right, there is a second figure lying on the ground very close to the first figure. That's a little tidbit that you wont hear from news agencies around the world, but if you look at the original NASA photograph, you will see the first figure is actually looking down from the rock its sitting on, as it seems to glare at the second figure lying on the ground. When looking at the original photo, I was surprised that news agencies laughed and joked about the figure. The skeptics blew off this photo as wind shaped rock mixed with the vivid imaginations of the public. Perhaps that caused less people to actually look into the photograph itself for fear of being ridiculed or embarrassed in front of the scientific community. That is called disinformation, when the governments release (using media as their proxy) wrong information to counter the real UFO information in order to confuse the facts.

I took the original photograph and enlarged it several times using Windows Picture and Fax viewer. Then I began scanning over it carefully trying to find other figures...success and I found it so close to the first figure that it could not possibly be a coincidence. It was just a few steps in front of the first figure.

Next I began scanning over the rest of the photo, which was a large job since it covers so much area of the Martian ground close up. I found a sculpted reptilian like face that was laying face up on the ground (Waring 1). Compared in size to the figures, this face was obviously carved from stone and its size was three to five time longer than the total height of the standing female figure. This face is on the right side of the figures, or left quarter, center of photo. We get a clear, detailed side view of it. Its mouth lines and small black eye make it appear reptilian in appearance, yet on top of its head it seems to have a headdress. (Note, the dark tinted color of the eye seems to confirm reports of the coloration of alien eyes seen in contact cases.) The headdress is similar to that of an Indian chief with the feathers standing out, only the face here does not have the hat go below the ear. The ear looks like it goes in a 30-40 degree triangle and reach a sharp point that is 1/3 the distance from its forehead to its chin. The ear goes out straight, similar to a cat's ears laid back when they are irritated or angry. This stone face appears to be attached to a neck and shoulders, mostly buried under the sandy dirt. If it were an entire statue with most the body buried under the sand, then its size if it were standing up, compared to the figures would be close to that of the Statue of Liberty. The actual size of the figures is still under question.

Where there's smoke, there's fire I always say, so I went looking for other familiar things in the photograph. Then I found the third unusual anomaly in the NASA photo. In the center of the photo was what appears to be not just one, but three faces. I named them Waring 2. The three faces are all next to one another, but with a single empty space between that would fit one more head into it. The three faces are each of a different species. One looks lizard-like (reptilian) with unusual extending skin like eyebrows and a flat head, the other looks buried to the chin and has a face similar to a coyote, yet flatter in appears. The third face is unmistakably a face and not a wind weathered rock. It is the largest of the faces and is set above the others on a slope. It appears to be semi-human-like, yet is more similar to chimpanzees than humans. Three faces, looking the same direction, on the same slope, next to one another would eliminate most skeptics and non-believers about the existence of aliens on Mars. Why they were built I can only speculate that it may have been created to commemorate three great leaders of three space-faring species coming together for a common good. I SCW will release several videos of these and other objects in this Mars photo on most video sites including Youtube.com. So if you want to see it go to Youtube and type "TaiwanSCW," that's me.

It is clearly impossible for NASA to cover up all objects that resemble man-made structures using NASA's version of Photoshop-like program to blur objects, but it is obvious that NASA will try to keep up the charade by hiding as much evidence as they can before its seen on their web sites. Many stories have come out of NASA, about people seeing photographs of entire cities many times larger than New York City. Those cities and structures have been seen on not one, but all planets and most moons, but the eyewitnesses are usually under contract not to tell anything during their employment with the government or they would suffer the repercussions deserving of revealing such secrets. Rumor has it that the US government is on a 200 year step by step plan, integrating alien technology along the way, at the end of which they will reveal aliens to the world, and most of all, the real reason why they hid it for so long, but NASA is 50 years old, so I assume we only have 150 more years to wait. But from the looks of it, it may be in the next 15 years if the public has anything to say about it.

NASA's Mars Exploration Rover Spirit took the photo itself that caused all the controversy, while it was looking at the westward view from atop a small plateau. The Spirit Rover rested on that plateau for several months of 2007. The plateau is called "Home Plate," and sits on the inner basin of the Columbia Hills range within Gusev Crater. Spirit used its panoramic camera (Pancam) to take this photograph somewhere between November 6th and November 9th of 2007. Spirit landed on Mars January 4, 2004, making its landing site somewhere on the center horizon of this very photograph.

Amazingly enough, the rover has continued to function effectively seventeen times past its expected mission end. As of December 2008 the Spirit rover is still hard at work and functioning well enough to keep it moving around, exploring and photographing as much as it could find. NASA currently has plans to launch a new rover onto mars in 2009, but budge cost will delay it a few years. This time it will be a super-sized, souped-up nuclear-powered rover and is called the Mars Science Lab. It will roam the plains of Mars in search for rocks that may contain microbial life. As if the figures on Mars were not large enough! It will have many powerful instruments that can probe rocks and soil in greater detail than the former rovers. This new rover will even contain a laser that can zap large rocks from a distance, possibly to break them and study the fragments.

Lets look at the publicized female figure in the photo from a psychological perspective for a minute. When you look at the main figure in the photograph, you notice some non-verbal cues from its position and stance. First off you see that it appears to have curly brownish hair that stops at the shoulders and to top it off, there seems to be a slight pinkish area in front of the hair, possibly being the persons face looking in the direction of the second laying down figure. Below the female figures right arm can be seen the curvature of the right breast. The shoulder can be clearly seen, leading to the elbow. The elbow it is bent at 20 degrees. From the elbow you can see the joints at the wrist and then the hand. The figures pinkish skin stands out from the dark green cloths the figure wears. At the point where the arm meets the wrist, there is yet another bend in a slightly downwards direction at about 30 degrees. From these angle changes from the shoulder to the hand, we can see that it has all the aspects (joints) expected of an actual human arm and hand. When you follow down her back along to her buttocks, the right upper thigh goes to the knee at a 25 degree angle downwards till it meets the knee, and at this point the lower leg goes backwards placing the foot below the figure in order to allow balance. The left leg is harder to see, yet the shadows from the green tunic it wears indicate that the left leg is extended almost straight out without bending. The green tunic covers from the neck down (neck not visible) and extends to stop at the elbows and at the ankles of the figure, but wraps around the figure not loosely like a Roman tunic might, but tightly except from the knee down. The female figure is waving to the other person to get down or to lower himself, possible to prevent them from being seen by the Spirit Rover.

Now we have seen with our own eyes that the figure has a head, shoulders, arms, elbow, wrist, hand, buttocks, upper and lower legs, knees, pink lower arms and hands and a pink face, not to mention beasts, so why do skeptics or NASA scientists still call this creature a rock? How many rocks like this one, created by nature, have you seen?

The second figure is within six arms lengths (first figure arm) of the standing figure. It lays down with its feet closer to the standing figure and its head furthest away from it. Its legs are together and are bent at a 20-degree angle from the buttocks area. A single arm seems to be extending down with the hand resting on the sand. The figures neck and head are clearly defined due to their pinkish tint as opposed to the dark green tunic it wears. The tunic again covers most the body with its right arm bent at a 45-degree angle and its left hidden on its other side. From the bend in its right arm, we can see that the person is turned towards the Spirit Rover. The tunic then stops slightly below the neck area, revealing part of the pinkish upper chest. It has little hair if at all, because the head appears to be bald, and very pink when enlarged. An interesting thing about these two figures is that both seem to have their heads turned in the direction of the Spirit Rover, as if they came to see this unusual object from Earth. Their green tight tunics stand out powerfully from the red-yellow dirt and rocks of their surroundings. Coincidence? Perhaps, yet how many coincidences have you heard about of similar rocks on our planet? I for one have never seen anything that can remotely be compared to this photograph, unless I compare it to photos of a human family.

The clothing appears to be tunic-like, yet has some unusual qualities that one might not expect, along with some subtle difference from the standing figure and the laying down figure. Both tunics seem to have a dark green color, yet they also have a shiny appearance or reflectiveness. Both figures have similar in design clothing so perhaps the cloths themselves are made of some special material that reflects off the heat, keeping their bodies cool in the Mars heat. Also both figures have clothing that is near skintight. The standing figure has a tunic that covers her neck and when the tunic reaches her knee area it begins to loosen and spread out like a dress, stopping at her ankles. This is another indication that this figure is female, not male. The second figure laying down has a different design of tunic. It fits tightly over both legs in a pants-like appearance making it a higher probability of this person being male. Unlike the standing figure, this one's arms are covered by the tunic, leaving only his hands and bald pink head exposed. (Note, the pink is very pink, not light pink, so either the light reflected off a sweaty or oily face making it appear more pink or this person has unusually pink skin. This may indicate a lack of living on the surface, but instead of living below the surface in artificial light.)

This leads us to wonder, are they living figures or statues left long ago by an ancient culture? Well, lets look at it from this perspective; the figures appear to be looking in the direction of the Spirit Rover. Enlarging the photograph until the figure is large enough to be seen and using a DVD camera looking at the enlarged photo on a flat computer screen, to look at the figures and re-digitalize them, making it clear again (amateur CSI technique) allows us to make these assumptions more easily. Since there are four possible directions (North, South, East, West) that the figures could be looking, that gives us a one if four chance or 25% chance that a statue would be facing the direction of the rover and a 75% chance that they would not be looking it its direction. So now we have a 75% chance that it's not statue, and a 75% chance that it is actual human figures. That means odds are in favor that these figures are not rocks and not statues either, but are living, breathing beings.

The figures odd green tight tunics leave us wondering if the children stories about little green men from Mars, are not just stories after all, or sightings long ago on earth, handed down through the generations by word of mouth as was done before pen and paper. James Baldwin (1924-1987) once stated, "Every legend, moreover, contains its residuum of truth, and the root function of language is to control the universe by describing it."

There has been a rumor that the Spirit Rover has taken another photograph of that same area, but was a few days latter and the figures were found to be gone. This was supposedly leaked from a NASA employee to the public in an Internet blog, but I could not find it or confirm it. It may be possible for someone to find photographs of the same location, but from a different perspective, so that it could be examined more carefully, but I will leave that to you to follow if you are interested.

The London Times reported, "NASA scientists have been puzzled by the peculiarly life-like image." NASA scientists then stated that the figure in the foreground was actually only two meters away from the Spirit Rover when it took the photograph and that this would put the size of the standing figure to be three centimeters tall. If this is true, then the rover has been driving around Mars, possibly running over houses and cities or crushing human-like life forms like Godzilla stomping through Tokyo. On the other hand, NASA is not known for telling the public the truth when it comes to aliens, alien structures or even alien ships that were clearly found. It may be that they said this to try to damper the world excitement over this photograph, hoping that it would fade into obscurity with NASA's other blunders.

The Spirit Rover's number one mission overall is to find if there is life on Mars, yet they shrugged their shoulders on this one and called it a rock without showing another photography closer up of the figures, which they do have in their possession. Also NASA has photographic and computer equipment that can re-digitalize and focus the figures 100%, yet again, NASA did not show the focused version of these figures. Although NASA holds all their photographs for six months or more before they release it, evidently this particular photograph slipped through without them retouching it. Anyone who has ever looked at Mars photographs has seen photos that are 100% clear, which have small oval like blurred areas in it. Most photographs have several if not many. These blurs may only be 2-5 millimeters in size or larger depending what they are covering up and its distance from the camera.

In 2009 it was discovered that 30% of Mars was suddenly covered in Methane Gas. This gas NASA said, came from below the surface of Mars. Understand that Methane is an unstable gas, which can survive only a few hundred years, yet it was just released in a few months over 30% of the planet. Methane is considered one of the four best candidates for detecting habitable conditions for life. Science dictates that a large release of methane gas would need a source that is replenishing, or continuing. This makes us believe that life in some form exists below the surface of Mars. All living creatures expel methane, this is especially true at NASA. It's a clear sign of life.

In conclusion, the Spirit Rover clearly found life and not just any life, but human-like life living on Mars at this very moment. NASA is suppose to turn over all their information to the public that they gather, but it is obvious, and evidence backs this up that they have been retouching photographs. That NASA has found life on Mars and is keeping the information from getting to the general public. NASA has stepped out of bounds and now dictates to the public what it will and will not allow them to view and know. NASA is an agency of the United States Government, responsible for the nation's public space program, with civilian and military aerospace research and development falling into that area. Back in 2002, NASA's mission statement read, "To understand and protect our home planet; to explore the universe and search for life; to inspire the next generation of explorers...as only NASA can." Strangely enough in February of 2006, NASA deleted this part of the mission statement, "to understand and protect our home planet." The reason for this is evident; that statement means that they already found life on other planets, raising the possibility that another species or race might attack. The technology in the public sectors is getting better and better, putting Joe Public closer and closer to getting his hands on re-digitalizing programs and equipment. That should be in about four more years, or in 2012 before the truth come out. Yep, you heard me right, 2012.

Chapter Eighteen: Needles, California UFO Crash

In the early morning hours of May 14th, 2008, multiple eyewitnesses saw a large glowing object in the sky. The object in the night sky had a turquoise hue surrounding and trailing behind it as it fell near the city of needles, California, just south of Las Vegas, Nevada. Numerous eyewitnesses swore that it was not a meteorite that fell, and apparently the US military thought similarly, because a lot of US military helicopters came searching for the crash site and were seen hauling it away.

38. Photo of 1997 UFO sighting in Columbus, Ohio (similar to Needles sighting). (At http://scwbook. blogspot.com/).

39. Photo of Janet Plane. (At http://scwbook.blogspot. com/).

40. Photo of Sikorsky S-64 Skycrane. (At http://scwbook. blogspot.com/).

It was three o'clock in the morning when many residence of the area saw and heard the UFO crash take place. Fifteen minutes after the crash, the sudden appearance of many unmarked vehicles with government license as well as workers in plain cloths. Five military helicopters were seen, including one Sikorsky S-64 Skycrane that was used in picking up and transporting the crashed UFO. The government like always, tried to cover up the incident as fast as possible, in order to stop the general public from finding out and possibly panicking.

A major witness of the May 14th, 2008 UFO crash was a person that was referred to in the news as only "Bob." This person lives on a houseboat in Topock, Arizona. This tiny town is so small it only consists of a Mariana for boats and a few stores. It sits just a few miles Southeast of Needles. Authorities in the area said that Bob reported seeing an incredibly powerful glowing object high above him in the sky that was heading in his direction. He said the object had a turquoise-blue-green light around it and a self-luminescent trail behind it. It all began at three o'clock in the morning. He was lounging out on the upper deck of his houseboat when he spotted the strange craft coming down from the sky. He watched it as it passed. It appeared that the craft was consumed in flames. He saw the blue object crash into the ground at a 45-degree angle somewhere within 100 yards West of the Colorado River on the California side. He was close enough to actually see it hit the ground and watched it bounce once. Then he heard a loud noise that sounded like a thump, but there was no explosion.

It occurred to Bob that it may have been some kind of plane that he had just witnessed, so he picked up his satellite phone to immediately call 911. This phone which he said never in many years had any trouble to work, suddenly would not connect (Remember, a satellite phone is unlike a cell phone in the way that it is often used by reporters who need an instant and guaranteed connection directly to satellites anywhere in the world. They never fail). Then Bob began moving his boat out into the river to get a better connection, which was when he heard the humming of many helicopters approaching. He stood on his houseboat, still staring up at the night sky, when he first spotted five military style helicopters flying in formation just sixteen to seventeen minutes after the unusual craft struck the ground. One of the helicopters broke away from the formation and began circling Bob's houseboat with searchlights scanning over the surface of the water and then the houseboat. He said the helicopter then returned back to the formation again. One of the helicopters was clearly different from the others. Bob recognized it right away as a military painted Skycrane that could carry and transport objects about the size of a semi-trailer. The Skycrane was used in retrieving the crashed object, which Bob watched and reported as being oval-shaped and still glowing blue as the Skycrane flew away with it.

Bob stated, "It was about the size of a semi-trailer. An oblong shaped thing." It had been picked up using some cables or a net and the Skycrane quickly flew away out of Bob's line of sight.

The second eyewitness to the UFO crash was Frank Costigan, who lives three miles east of the Colorado River. He had gotten up at three in the morning to let his cat outside, when something caught his attention. As he was walking around his back yard, Frank watched a large glowing craft shooting down from the sky. As the object traveled it seemed to flash several colors that included turquoise, blue and green. The object was traveling towards him at incredible speeds. When he saw it crash to earth in the distance Frank was expecting to hear some kind of impact noise, but he was surprised to hear nothing.

Costigan stated: "I thought I might hear something when it hit the ground, because if it was as close as I thought it was, and as big as it was, I thought I would hear something, but I didn't hear anything. And it went out of my view before it hit the ground. It went behind a hill and I waited to see if I could hear it crash, because as big as it was, it was bound to make noise."

Frank Costigan said that as the object got lower and lower, it illuminated the ground. It came from the northeast and was going in a southwesterly direction at incredible speeds, slowing down, then speeding up again, and soon leaving his line of sight. Frank is a retired superintendent of operations and police chief for the area. Frank formally worked at Los Angeles International Airport (LAX) from 1978 until he quit in 1985 to work at the Ontatio (California) International Airport from

1985 until 1986. His past working experience made him very aware of what typical aircraft can do and generally look like in all weather conditions, so his account of the UFO crash is making him the most important witness of the entire event. He sometimes will do some special reports and news investigations for an am radio station in Needles, California called KTOX. When Frank told the radio station owner David Hayes, about the remarkable sighting that he had in the early morning hours, Hayes responded with his own version of what strange things that he had seen that day.

Just a few hours after the UFO crash, when Hayes was driving along I-40 making his way to work that morning, he spotted a small convoy of darkly painted vehicles exiting the highway. The strange vehicles had black and white government license plates. The truck that was evidently leading the convoy looked like a really large SUV with a dome on the top and a truck bed in back. It clearly had four-wheel drive. He believed that the vehicle was carrying a remote controlled drone on its bed. He described seeing a triangular shaped object that sat on top of several smaller humps in the truck bed and it reminded him of the appearance of a stealth bomber with its wings folded. (Perhaps the drone was used to find the location of the falling blue UFO the night before). Behind the truck was a dark-green van following closely. Hayes told Costigan that he got a good look at the men inside the trucks and it seemed to him that they had the behavior and focus of military. He saw that they did not have on military uniforms, but they did notice him when he made eye contact with the driver of the third vehicle. Strangely enough, that third vehicle left the convoy and followed Hayes to the radio station. Later that same day, one of the trucks from the convoy returned again and parked outside the KTOX building for a few minutes. It seemed to be making a security surveillance of the radio station. Hayes said that the surveillance had a sort of men in black (MIB) appearance to it all. In his words, "They were as serious as a heart attack."

Hayes said, "The fact that there were people here the next day, it was almost like they were doing some sort of cleanup or whatever. The point is, something definitely happened."

This news was too incredible to keep from the general public, so Hayes and Costigan decided to do a radio show on it. During the radio broadcast they asked for any witnesses of the May 14th, 2008 sighting to please come forward to help shed more light on what happened that night. That is exactly how the information from the person only known as "Bob" came to light, but he was not the only person to call into the show. A friend of theirs from Laughlin, Nevada, called into the show to tell about his experiences of seeing Janet planes taking off and landing at the Laughlin Airport all night long on the very date of the UFO sighting. Hayes says that he has known the eyewitness for years, and was often referred to as Bob...Bob on the river.

For those unfamiliar with Janet planes, they are the nicknames given to the aircraft that often fly workers in and out of Area-51 and Area S4. Remember, Area S4 is 13.5 miles South West of Area-51, yet S4 is where they take recovered alien technology and back-engineer it to find out how it works and how they can make it themselves. Just like all airports that have Janet planes enter and depart, the Laughlin airport and control tower reports that they were closed at that time, so airport personnel could not confirm the presence of the Janet planes on the night in question. This is a typical response when Janet planes are present anywhere. They take non-military scientists from their city and land directly at the Air Force Base that goes by the many aliases of Groom Lake, Area 51 and Dreamland. Note, Area S4 also has a small runway for smaller prop driven airplanes. Also note that most of Area S4 sits 3-4 miles underground, below a tiny airplane runway, with only a handful of visible aircraft buildings around, but all in all, there was no reason for the Janet planes to land in Laughlin Airport. (Note: Using Google Earth, look in Area S4 for Lazar's

Hanger along the mountain ledge on the left hillside. Holds eight different UFOs).

More unusual events seemed to continue after the sighting of the UFO crash in Needles that morning. It seems that no one has been able to find and locate the man only known as Bob, causing some to believe that he has left the area to avoid US Air Force personnel in plain cloths coming to visit him and others believe that after the radio show, the person known as Bob may have been found by the government, only to disappear...forever, much as some witnesses disappeared at the Roswell UFO crash of July 1947 at Roswell, New Mexico (Note: panicking nurse from alien autopsy).

Toni Sagan who is a member of the River Valley Democratic Club in the local area was surprised when she was questioned by a stranger at one of the club meetings, right after she appeared on the radio station where she talked about the 2008 presidential race. The man was very interested in any off the air comments between David Hayes and an eyewitness named Bob that had phoned into the show. Toni told him she didn't know anything about any of that, and she noticed this man seemed out of place with their comfortably dressed community from the way he acted and dressed.

There has been a big media blockade that has stopped information from getting out about the UFO crash to the general public. The police, emergency, military and government agencies are unwilling to contribute information regarding the crash. CBS investigative reporter George Knapp and members of the Channel 8 (CBS Network TV Affiliate in Las Vegas) Investigative Team said that they have contacted "police agencies in three states including the Mohave Tribal Police, the Laughlin Bullhead International Airport, the National Weather Service, the FAA and Nellis Air Force Base and China Lake Navel Base" asking for any kind of statement about the UFO incident that took place on May 14th 2008. All of which are near the UFO crash site. Every single one of them said that they knew nothing about the UFO crash. A military watchdog group was lucky enough to find a public record showing that there was at least a single army helicopter in the air, in that exact area, at that exact time. The helicopter leads us to some new evidence. The helicopter was not where it was registered to actually be, but is instead listed as being attached to a U.S. Base in Europe. (Perhaps this is how black opps accumulate their hardware and hides their existence, by borrowing equipment and supplies from other US military locations).

We know that based upon the forty-five degree trajectory of the falling object, that had it been a meteorite, its trajectory would have caused it to most certainly burn up in the atmosphere, because the trajectory caused it to travel through atmosphere longer than usual for a meteorite. Most meteorites burn up in the atmosphere due to this fact. It is possible it was space junk, but the fact that it changed speeds from fast to slow, then fast again, tells us that it was under intelligent control. Although it is possible that it was a top secret aircraft referred to as Project Aurora, because of its glowing description and its sudden slowing down and also apparently being salvaged still in one piece, although those speeds reported should have obliterated any known craft, stealth or otherwise.

This leaves us with only three final possibilities. First, it was being tracked by the US government and was shot down when it got close enough, so that the government could get their grips on another alien UFO. Secondly, the UFO was one of the alien crafts at Area S4 flown by human pilots currently trying to learn to not just back engineer, but to test fly the craft in and out of Earths atmosphere, or even to a planet who's space-faring inhabitants may have traded the US Air Force the UFOs in exchange for something of value to them. Third, the possibility that the US military is communicating with and inviting alien species in hopes of gaining a technological advantage over other countries.

As a side note, in 2009 the United Kingdom release a slew of UFO documents, some of which include information mentioning the USA military. One of which was a UFO sighting in the UK back in 1993, but UK government investigators came to the conclusion that it was not an alien controlled craft, but instead was a secret USA military project called, "Project Aurora." The term Aurora is used because the craft causes an aurora-like greenish-blue glowing affect. One report about it mentions UK police witnessed two objects in the sky seeing, "two vapor like trails appearing behind each object and they appeared to be self luminous." (Case No. 933, March 31st, 1993 at http://ufos.nationalarchives.gov.uk/). Perhaps directly linking it to the current UFO crash in Needles in May of 2008. The similarities are uncanny.

I have to admit, of all the UFO sightings that I have researched, this one hit me on a more personal note. I, SCW once lived in the small city of Needles for three years when I was growing up. This is a place with such a clear sky every night, that you can see a shooting star every 10-20 seconds and summer days so hot you can cook an egg on the sidewalks or the hood of your car, not to mention during lighting storms, you would see lighting bolts of every color imaginable.

The extraordinary city of Needles is located in the desert in the middle of nowhere, along the southern boarder of California in the Mojave Valley. The city is only accessible using Interstate 40 and U.S. Route 95. Beyond the city of Needles is only rocky desert with tumbleweeds and a lot of needle like mountains that gave the little city its famous name. In 2008, the population was estimated to be 5,700. The temperatures can often reach 120 degrees Fahrenheit or forty-nine degrees Celsius, but lets get back on subject.

The Sikorsky S-64 Skycrane is an American turbo shaft twin-engine heavy-lift helicopter that has a six-blade rotor. The S-64 has a two-person crew, but can hold up to five. It can carry a payload weight of 21,000 lbs (9,072 kg). It can travel at a maximum speed of up to 126 mph or 203 km/h, but generally cruises at speeds of 105 mph or 169 km/h. This helicopter can carry loads connected to long cables or slings in order to pick up and carry heavy loads without having to ever land. Its maximum range is 400 km or 250 miles. This means that the base it came from and went back to, had to be located 125 miles away or closer. Let's say that the helicopters were traveling at 105 mph, since it took a reported 16 minutes to reach the UFO crash location, we can deduce that the S-64 travels at 1.75 miles per minute. Therefore the S-64 Skycrane had been approximately 9.14 miles away from the crash site, assuming that the helicopters were already on standby awaiting the crash. The Needles Airport is approximately 8.35 miles Southeasterly of Topock, Arizona, where Bob's house boat was at the time of the crash, therefore we now know the helicopters temporary base of operations that morning was at or near the Needles Airport (8 miles South of Needles) not the Laughlin Airport (20 miles North of Needles) where the Janet planes were seen.

The Needles airport has only two short runways for planes to land on. Each is approximately .75 of a mile long. This is too short of a runway for Janet planes. There is a single building with a single hanger next to it, so for a 727 to land here, when it was shut down would have been a difficult task considering the short runway, however the military helicopters had no trouble using it. The Laughlin Airport on the other hand, has a runway that is 1.6 miles long, making it much easier for two 727 jets to land.

The city of Topock is located 178.6 miles Southeast of Area 51, home of the Janet Planes, and home of Area S4, the area to study alien technology (Yes, Area S4 is real, as is Area 51). Area S4 is the place scientist Bob Lazar says he saw eight real UFOs, one of which he worked on trying to figure out its three engine propulsion system by extracting one of the three and taking it apart, and back engineering it. Area S4 is 13.5 miles southwest from Area 51 main airport landing strip. Just check it out on Google Earth map (type, Area 51) and see for yourself, or Google search or on Youtube.com. (Type, Bob Robert Lazar, Area S4).

UFOs like this one have been seen repeatedly in the Needles, California area over the past several years. An eyewitness to such a UFO had been driving their automobile heading West of Needles on I-40 on April 17th, 2002 saw a similar craft. The eyewitness at that time said it was three thirty in the morning when the event took place, (Note similarity: the 2008 UFO sighting was at 3 am). The eyewitness said that the UFO hovered over the very road that he was traveling upon.

In 1997 near Columbus, Ohio, a photographer who happened to be filming clouds, to his dismay saw a glowing object similar in description as the one in Needles. He caught the craft in several frames of the video, which shows clearly a turquoise oval shaped UFO over his home. The photographer said the strange thing was that at the time of filming, he had not seen a UFO in the clouds and had no idea that it had been there until he had gotten home to look at the video. Perhaps this shows that UFO hunters should photograph the clouds using high megapixel digital cameras (14+MP), which may yield them better results.

The bluish-green UFO that crashed that morning near Needles, California is an overwhelming fact. There were eyewitnesses so close, that the US government investigated them. The fact that they followed the radio announcer to his radio station (KTOX) only shows how much the USAF wants to keep this case close. I'm sure in the future we will hear more about this bluish-green UFO being seen in various places. Its evident that the classified documents released by the UK that revealed 'Project Aurora' will gradually bring the truth to light, but until the US releases such documents, we remain on our own to find the truth. Although this UFO crash appears alien in nature, I do feel that it is the USAF testing their newest alien technology. For this craft to fall at such a speed and be unable to slow down dramatically indicates that it came from a very far away distance and the pilot clearly has a lot to learn about controlling the craft, as evident in its rough landing.

Chapter Nineteen: UFO Strikes Windmill, January 4, 2009

On January 4, 2009 in the small town of Conisholme, in the eastern part of England, something crashed into a windmill at 4 AM, leaving a lot of speculation on how it happened. Ecotricity is the company that owns the wind turbines. Many engineers from Ecotricity have been working on some solutions of how this could have happened, yet with all their vast knowledge of wind turbines, they came up with no answers. The local residents however claim to have seen UFO lights flying around that night, which they are certain led to the destruction of one of their windmills. One blade was torn right off. The other two blades were still attached, but one of them was bent in several places from hitting something hard. The owners of the wind turbine farm searched everywhere for the blade that fell off, but it was hard to find. Latter the workers located it. Even the UK newspaper, The Sun, placed the broken wind turbine on the front of their newspaper with the headlines, "UFO Hits Wind Turbine."

41. Photo of windmill hit by UFO. (At http://scwbook. blogspot.com/).

42. Photo of Aurora Cemetery. (At http://scwbook. blogspot.com/).

The windmill tower itself is a whopping 89 meters high, making it near impossible for someone on the ground to knock it off. Each blade is 20 meters in length making its total reach on either side at 40 meters across as it makes its rotation. The main part of the tower is constructed using fiberglass. Any lighting strikes would melt into it leaving burnt marks. These windmills are made to withstand extreme conditions, but evidently this one met something that it was not prepared to withstand.

Although lighting was first believed to be the cause, no burn or scorch marks were found. The white paint of the blades and tower remained like new, without burns. If by chance lighting did hit the blades, it would not have cut it off, but instead would leave most of the blade intact, although some melting might occur.

Local researcher Russ Kellett, of the Flying Saucer Bureau stated, "This is the most reports' (of UFO) activity we have ever had. I have had over 30 phone calls and emails. For it to hit two of the blades, the object must have been about 170 feet long." He is referring to some kind of flying craft hitting the first blade, then continuing until the second blade hit it.

The second blade was bent in the middle at its strongest point, yet lighting could never cause such damage.

Some of the local residents have claimed to see some strange flashing tentacle shaped lights hovering above the windmills on the night of the occurrence.

Local news agencies reported that the weather in the area was normal and the wind was calm that night and morning.

One local resident that witnessed some unusual lights was John Harrison from nearby Saltfleetby. He spoke about how on January 3rd, 2009 he was looking out his landing window when he noticed a massive ball of light with tentacles going right down to the ground over Conisholme Wind Farm. He stated: "It was huge." At first I thought it must have been a hole where the moon was shining through, but then I saw the tentacles.

It looked just like an octopus." (Note, he said the tentacles actually went down, touching the ground. The craft needed contact with the ground for soil testing or communication to an underground base? This may indicate that the windmills were not interesting to them, but were actually in their way).

Residents remember being woken up early on Sunday morning at 4 am by a huge crash. Latter that morning they discovered one of the wind turbines was bent and one of its blades had gone missing (but latter recovered a long ways away). Several of the witnesses reported that they had seen glowing spheres in the sky above the Wind farm just 8-6 hours before the crash.

One woman stated: "I just suddenly saw this light. A pair in front of us and it just seemed to whiz sort of, across the sky... toward the...wind turbines." Then she went on. "There were two or three of us in the car and I just said to the other two, what was that?"

A Gentleman standing beside the woman stated, "I did see lights in the sky, yeah, very low. You know, so it would be at the height of the turbines when it got there, if it was something like that. A UFO or something."

Dale Vincent, the founder of the company Ecotricity stated; "a lot of people were there on Sunday. We had the manufacturers team there the next day and they have been crawling all over it and the turbine, looking at the damaged parts, and we shipped some parts off for some forensic testing to determine better if we can, what the probable cause was at the moment. We are left with no kind of firm favorite to what has happened. The blade was later found and sent off to Germany to be tested, yet no results were made public."

In a BBC report, Mr. Vincent says, "It's a very interesting story actually." Then he went on to say, "We have been onsite now for a few days trying to find the cause. Funny as it may sound, the UFO theory is the best one we got." BBC went on to say, "Something big and heavy hit this thing." With both the

BBC and Mr. Vincent believing that something flying had to have hit this windmill, it further presses the issue of why was a UFO over the wind turbine farm? Perhaps investigating Earths' alternative energy sources?

This sounds logical, but if one of the UFOs crashed into the turbine, then it can also be assumed that they were not studying it. Had they been studying it, they would have been watching it too carefully to crash into it, but instead they may have been interested in something below the ground, with the turbines in their way.

Some speculation suggests that it was the buildup of ice that may have caused the blade to snap off, yet a thorough search of the area could not turn up the missing blade until much later and strangely far from its original location. At first local officials were confused by the missing blade and had no idea what could have caused this to happen, but the locals seemed to be confident that the UFO sightings that many residents had seen the night before, had to be the cause.

So why on this night and early morning would a UFO be flying around the area at 80-100 meters above the ground? Note, that it was a wind farm that the UFOs were seen over, so this gives us several logical reasons. First, the UFOs were stealing energy from the wind turbines. This doesn't pan out since if that were the case, UFOs could go anywhere to suck power lines or power plants, let alone the fact that their technology would be so advanced that they would have their own endless supply of power. The second reason seems more logical. They were just there to map out and record information about how humans are trying to harvest the energy from the wind. The fact that humans are trying to find more eco-friendly power solutions may interest them. If this is true, it is nice that aliens are aware that humans have started a going green movement around the world in an attempt to save our species from extinction. Clearly this attempt intrigues them. I say human extinction rather than saving the planet, because the planet will always be here, yet may not always be suitable for life. Third possibility is that they were trying to communicate with an underground base, but since this seems unlikely, since this was only one sighting, not many over the year in the same location.

Ridiculous some might say, but let me remind you about another similar UFO incident that took place long ago and was recorded in detail in the local US newspaper. On April 17, 1897 in the city of Aurora, Texas, a slow moving UFO had flown into and crashed leaving a single tiny human-like body. Also some of the debris had engravings of hieroglyphic-like symbols. When the body of the tiny creature was found at the crash site, the local people felt sad for it and decided to give it a proper burial. The burial took place at the Aurora Cemetery where the grave can still be found today with a simple marker.

An account of the UFO sighting was reported in the April 19, 1897 edition of the Dallas Morning News. Aurora resident S.E. Haydon wrote the story. In the story they reported that the cigar-shaped craft hit a windmill on the property of Judge J.S. Proctor two days earlier on the 17th at about 6 AM. This impact with the windmill resulted in the crash of the UFO. The story continued saying that the pilot was "not of this world," and the paper went on to say that an Army officer from nearby Fort Worth saw the body and called it a Martian.

The Dallas Morning News stated this on April 19, 1897:

"About 6 o'clock this morning the early risers of Aurora were astonished at the sudden appearance of the airship which has been sailing around the country.

It was traveling due north and much nearer the earth than before. Evidently some of the machinery was out of order, for it was making a speed of only ten or twelve miles an hour, and gradually settling toward the earth. It sailed over the public square and when it reached the north part of town it collided with the tower of Judge Proctor's windmill and went into pieces with a terrific explosion, scattering debris over several acres of ground, wrecking the windmill and water tank and destroying the judge's flower garden.

The pilot of the ship is supposed to have been the only one aboard and, while his remains were badly disfigured, enough of the original has been picked up to show that he was not an inhabitant of this world.

Mr. T.J. Weems, the United States signal service officer at the place and authority on astronomy, gives it as his opinion that he was a native of the planet Mars.
Papers found on his person-evidently the record of his travels-are written in some unknown hieroglyphics, and cannot be deciphered.

The ship was too badly wrecked to form any conclusion as to its construction of motive power. It was built of an unknown metal, resembling somewhat a mixture of aluminum and silver, and it must have weighted several tons.

The town is full of people today who are viewing the wreck and gathering specimens of the strange metal from the debris. The pilot's funeral will take place at noon tomorrow.

S.E. Haydon."

The unusual thing about the Texas sighting was that the newspaper article written by S.E. Haydon was so casual and so forward with information about everything that obviously the US Government had not yet decided to keep UFO knowledge a secret. The simplicity and information given in the article make this sighting one of the most straightforward and honest recorded sightings in history. We can only imagine the things that S.E. Haydon had seen and held in his hands.

This leads me to more questions like what happened to the written records of the pilot that no one could read? What happened to the metal that was moved around? Was there no one with at least a single fragment of this craft still? Why hasn't the grave of the alien been dug up so that further examination could take place? It seems that answers only create more questions.

Wreckage from the crash of the UFO was later dumped into a nearby well located below the damaged windmill. Some of the wreckage is also reported to have been placed into the grave of the tiny human-like pilot. In 1945 Mr. Brawley Oates purchased the property from Judge Proctor and cleaned out the debris from the well so that he could use it as a water source. Latter Mr. Oates developed extremely severe arthritis. He claimed that it was the result of the contaminated water from the wreckage that had been dumped in the well. After this, Mr. Oates sealed up the well with a concrete slab and placed an outbuilding on top of the slab. This is according to the inscribed writing upon the cement slab itself, which was discovered in 1957.

In November 19, 2008, UFO Hunters TV show did an episode covering the Aurora crash. In the well they found that there was a large concentration of aluminum present, and all metal fragments were gone. I do however feel that they should have been looking smaller fragments, rather than larger ones.

In conclusion, many disbelievers might say that if UFOs exist, they would be too sophisticated of craft to crash into a windmill. The Aurora crash in 1897 would prove them wrong. The small town of Conisholme, England on January 4, 2009, had many people already say that they saw a lot of unusual lights and objects flying over their small town the night before the crash. Something hit that wind turbine, and hit it hard without crashing, making this UFO much larger in size than the small single person craft of the 1897 Aurora crash. As Russ Kellett stated, "For it to hit two of the blades, the object must have been about 170 feet long." Perhaps aliens might be curious at how humans are changing their ways of thinking about energy sources, and upon seeing the wind turbines, they might see that humans are trying to preserve the environment, rather than continue in its destruction.

Chapter Twenty: 2009 Ring UFO over Kings Dominion, Virginia.

On June 13, 2009 WAVY news in Virginia aired a news broadcast that showed a large dark ring in the sky above the Kings Dominion Amusement Park. This mysterious circle floated over the theme park making a lot of eyewitnesses in Kings Dominion, Virginia wonder what it could possibly be? Not just one, but also hundreds and maybe thousands of spectators at the amusement park were eyewitnesses to this black ring below the clouds.

43. Photo of Virginia 2009 UFO ring. (At http://scwbook.blogspot.com/).

44. Photo of Fort Belvoir, 1957 UFO. (At http://scwbook.blogspot.com/).

WAVY news stated that a woman and her sister contacted WAVY news saying that they had personally witnessed this strange object and asked the news station if they had any information about it. These two women from Hampton Road saw the ring and said it was a very moving experience.

A highly detailed video of the ring slowly moving through the clouds over Kings Dominion Amusement Park was recorded. The video was posted all over the Internet including Youtube.com. In the video the person behind the camera was focusing on an amusement ride. The ride is a giant pole that has a ring around it (coincidence?) and holds about 30 people around it. The ring rises up the pole until it reaches its top and then hesitates for a second before it drops down at high speeds. As the cameraman was looking up at the ride as it rose up the pole, off to the right was seen the dark ring in the clouds. The ring floated slowly through the sky and across the amusement park and out of sight, but did not appear to change speeds. Video of the UFO sighting can be seen at http://www.youtube.com/watch?v=O87h-fkTj1Y posted by user rpknowles on June 2009.

A newscaster on WAVY stated as he watched the video that it, "was certainly eerie to say the least."

When WAVY news contacted Kings Dominion Amusement Park officials, they said there was a simple explanation for the cause of the sighting. They claimed that it had to do with one of the rides at the park, which may have made a smoke-like ring. Eyewitnesses to the event say that it was too uniform to be smoke and that the ring never broke like smoke would. The amusement park believes, but is not certain, that it might be caused by a ride called the Volcano.

CNN interviewed an eyewitness to the event. Her name is Dina Smith and she is a writer. She said that she and her family were at Kings Dominion Amusement Park when they saw the black ring below the clouds. They all looked up and saw the ring. For the life of them, they could not figure out what it could be. Smith stated that she was confident that the ring was not smoke. She states in a CNN interview, "Smoke usually looks smoky, cloudy, a ring of smoke. This was a perfect circle. This thing was lined up so tight like it was a cut in the middle of the sky." Smith however doesn't believe it was a UFO. She stated, "It was like a sign. God gives you signs and I just feel like that was a sign and I'm not sure what that sign meant, but it meant a great deal to my family, because when we went home, we literally all got in line and prayed together." Latter on in the interview

Smith states that she still believes that the black ring is still out there somewhere.

Just when you think it was a single sighting of a black ring below the clouds, then comes the realization that this wasn't the first, but instead was one in a long line of black UFO ring sightings throughout history.

Sightings of this UFO black ring are actually more common than most would believe. There are actually hundreds of recorded and photographed black rings in existence on the web and in libraries. Its less likely that someone would deliberately try to make a ring UFO hoax photo or video due to the simple fact that it is a mere ring and doesn't appear like a spacecraft. Most hoaxers will try to make a saucer shape inspired from movies, but this is unlike regular saucers or triangles reported.

Take for instance the sighting that took place in 1999 at Orlando, Florida. In this sighting the eyewitnesses were lucky enough to get some really nice video of the UFO. In the short 17-second clip posted on Youtube.com, the video starts out with the person holding the camera standing in the street watching a blimp in the sky. Then the person realizes that there is a black ring to the upper right of the blimp. The black ring was clearly larger than the blip and since it was behind the blimp, which leaves me to speculate that the craft itself was so large that it could actually hold human like passengers within it. The ring craft undoubtedly had some sort of interest in the blimp that was passing under it at about 50-70 meters below the ring. The ring appeared to match the blimps speed as it can be seen traveling past power lines along the side of the street where it was filmed. An eyewitness that was leaving the golf course, noticed the blimp and began filming it. That was how this video came to be. The eyewitness stated that they were taping the blimp as it was traveling west to east in the sky, and said that they even made contact with the pilot of the blimp and asked him if he had seen the black ring. Sadly, the pilot said he did not see it.

This brings up another question, are eyewitnesses seeing a ring shaped craft or was the ring actually the outline edge of a clocked vessel observing the humans on the ground?

Another ring sightings happened near the Rocky Mountains in July 22, 2008. Two kids can be heard trying to focus the camera as a white-silver metallic ring sits below some dark black clouds. The odd thing about this is that the ring is standing on end, or sideways like a wheel in the sky. The nervousness in the kids voice sounds very authentic as an older boy tells the younger one to take a picture and the younger one says that he is. At that moment the craft accelerated from a hovering motionless position to a fast acceleration into the dark clouds above. (Note that there is little documentation to go with this video, however it does appear real and not a hoax).

Records go back as far as 1957, when in Fort Belvoir black and white photographs of the same black ring below the clouds were taken. To see the photos from the military, just Google Fort Belvoir, 1957, ring. The photos could easily be compared to the Kings Dominion, Virginia 2009 ring sighting. It seems beyond coincidence that the 1957 rings look very similar in size, color, altitude and position. During this sighting something unusual and almost creepy happened. A cloud was photographed that was round in shape, then photographed as the cloud began to dissipate, yet a dark line became visible. The photos continued to reveal that the dark ring began revealing itself from within its cloud-like shield until there was no cloud anywhere around it. It is obvious that the cloud was some sort of shield created by the UFO, because the time laps photography revealed that the cloud seemed to wrap perfectly around it and compactly around the edges of the ring.

With mass eyewitness encounters and video footage that is taken seriously by multiple news agencies, still the US government refuses to investigate such sightings and refuses to release information about them. With such a hard-core stance as that, it can be concluded that the US government has a lot of involvement in not just the sightings of UFOs, but also the cover up of the sightings. From this research, it can be assumed that the black ring does have a clocking ability that appears as a round or saucer shaped cloud. This ring that is visible also could actually be the outline of the ship, much like the gray-black outline seen in the O'Hare airport sighting in 2007. The ships can fly at low altitude and travel as slow as a blimp. How to deduce the truth you ask? Pick up your DVD camera or a 10+ megapixel camera (10 megapixel or higher picks up better details when photos are increased in size on a TV or computer), walk outside and take some random shots every day for a while. Then put them on your computer and enlarge them to see details. If you are lucky or pay a lot of attention to details, you may witness the next UFO sighting. Now pick up your camera and start observing the clouds please. Your sky is just as likely to have a UFO sighting as any place in the world, so start aiming for the clouds.

Chapter Twenty-one: Lighted Disk Shape Cloud Over Moscow, Russia: October 6, 2009

On Wednesday evening, of October 6, 2009, an odd shaped cloud appeared over the western part of Moscow, Russia. This cloud glowed with white light and was in the shape of a huge disk. It was so large that some who witnessed the event had said that it was similar in size to the mother ship in the ID4 movie. Along the edges of the cloud could be seen a white light, which shone brightly in the dark overcast evening sky. Understand this however; high above this white cloud were black clouds that totally blocked out the sky, therefore it is logical to assume that the light from the white cloud could not have come from the sun, since no sun could be seen through the separate sheet of black clouds that was far above the UFO. The video that has been seen all over the Internet and television news was shot on Moscow Ring Road in a car traveling from Volokolamsky to Novorizhskoe highway.

45. Photo of Moscow, Russia 2009 UFO cloud. (At http://scwbook.blogspot.com/).

46. Photo of Romania UFO cloud 2009. (At http://scwbook.blogspot.com/).

Like most UFO sightings, this one would have disappeared into obscurity if it had not been so brilliantly recorded on someone's cell phone camera and uploaded to Youtube.com for the public to take notice. Many weather scientists and specialists looked at the footage to find the cause of the white glowing cloud formation. They have ruled out chemical pollution and said that it could not possibly be any form of known weather phenomenon. Instead scientist say that it was no more than an optical illusion or in other words, your eyes fooled you into thinking you saw a glowing disk shape several kilometers in diameter. Yep…you heard it right. It's the Russian equivalent of the US responses of calling UFO sightings either weather balloons or Venus. You would think that someone with a doctorate in science could think up something better, but no… that pathetic attempt is the best they got.

You would think that a glowing cloud would cause the Russians to scramble their aircraft to check the formation out, but it may not have shown up on airport radar, but this is yet to be confirmed.

Lets talk about the video itself. At http://www.youtube.com/watch?v=nT2jwkCirP0 posted to Youtube.com by Russia Today a.k.a. RT News. RT is a TV channel in Russia that is government funded, but shapes its editorial policy free from political and commercial influence. The broadcast over six continents and 100 countries and can be found at www.RT.com.

Now this video reveals more than the newscaster at RT had known. First off, in the video there are other UFO or UFOs that were so small that most viewers would not notice. Let me help you to find them, but to see it, you must play the above video in full screen mode. At 15 seconds into the video, two dark black round tiny objects poke out from the far right of the cloud two times before they fly out at 16.0 seconds into this video. At the above video, stop the video at 16.0 seconds into it. The video itself is only 35 seconds total. Why so short? It appears to have been taken while driving, and we know how dangerous that could be. When you stop the video at 16 seconds, you will notice that there is a small black object in the far right inner edge of the cloud UFO. Now press play and you should see it. If not, continue doing this till you notice it. At 16.75 seconds into the video the object flies upwards out of the cloud and is seen above the UFO cloud in the upper right of the screen. The odd thing about this is that instead of one object, now there are two tiny oval shaped black objects. These are both easily seen against the dark cloudy sky. If you still cannot see the tiny black UFOs or probes, then press the full screen option to make it bigger. The last time I looked, this video had 578,648 views and for good reason, because its legit.

Another separate yet similar video is at http://www.youtube.com/watch?v=EopIdYAR5BE posted by 2012gregg and it allows the tiny black UFOs that leave the cloud to be seen more easily. The only discrepancy is that I can only make out one black probe-like object, where in the first video there were two. 2012gregg did notice the small object and writes that you should stop the video at 5 seconds into it and then move it second by second till you see the object. At 6.0 seconds into the video, a tiny black UFO can be seen either flying in or out of the UFO cloud. It is difficult in this video to assess if it's leaving or coming due to the tiny crafts incredible speed and low video quality. At 7.0 seconds, the object or objects poke out and back into the cloud at the upper right of the cloud UFO. I find this confusing since in the RT News video it appeared the reverse, so further analysis is clearly in order.

Russia is a country with a particularly large number of UFO sightings. Even in the old communist days, there was an overwhelming interest and in UFO investigations, although the official party rhetoric was to continue to publicly dismiss all UFO matters. For example in 1980, a worker in a factory upon the small Baltic island off the coast of Estonia was documented as being struck by an orange beam of light that shot out from a UFO that had landed nearby. He stated that the UFO's skin was "glittering with colors." Next to this craft were two standing black cubes with rotating pipe. The force of the beam was so powerful (unlike light) that it knocked him backwards. Long after the craft had flown away and disappeared, the eyewitness continued to complain of feeling nauseous.

Let's dive into one of Russia's most notorious UFO cloud sightings. It took place in 1974 at the Borosoglebsk Area, above Povorino Airfield. The airfield workers noticed a motionless black cloud that had appeared over the airfield. This black ominous cloud hovered at an altitude of seven kilometers or 4.349 miles up. The cloud was a kilometer and one half long or .932 miles long. The airfield radar indicated that the cloud was actually a craft. At this point a military jet was sent to intercept it with two pilots on board. As soon as the jet entered the cloud, an incredibly sharp and piercing siren pierced their helmet's earpieces. The sound was so great that it was above their pain threshold. At that same moment their onboard gauges illuminated alerting the pilots of being at a dangerous altitude. The aircraft began shuddering. The pilots somehow managed to shut down the jets power and desperately worked to get themselves out of the cloud. This black cloud hovered over the airfield for four hours before it suddenly disappeared. The Soviets were never able to determine what the cloud actually was, leaving a lot of guessing at to its origin and purpose.

Why would a UFO create a cloud formation that allows it to hide from view? The electrons coming off of an oscillating negatively charged self-accelerating object can knock out electrons in the atoms that compose our atmosphere. When the atoms regain their electrons, they will emit an aurora of greenish, bluish (Note: US Project Aurora), or yellowish colors. The smell of ozone may by noticeably detected if a person is at close proximity to the UFO. UFOs have often been seen making or leaving behind an aurora-like glow in the sky where it had once been. A ring of glowing air may also be seen along the circumferential edge of the UFOs. A small space between the outer metal surface and the ring is commonly found in UFO reports, which might be due to a reduction in air pressure along the smooth surface of the craft. When there is a reduction in air pressure and humidity, and there is humidity present in the atmosphere, clouds can easily form around the UFO. This is why I believe that some UFOs have been seen hiding within a cloud. It is possible that they do not mean to hide from us, but that it is an oddity of their UFO technology that causes it to appear so.

On October 25, 2009 in Romania, a similar yet different video was taken of a halo like white ring in the red clouds. The video of the event can be seen at http://www.youtube.com/watch?v=uWR_w7RqQPU where Romulo282838 was first to post the video. The video was taken with a cell phone camera, but this is not the same camera that took the UFO Moscow cloud, so I rule out that this is the same person trying to perpetrate a hoax. Instead it is an altogether different eyewitness recording a similar, yet slightly different sighting. The video was also taken from the front the drivers seat of a car, but the quality of the cell phone camera is much less than the first October 9th, 2009 Moscow sighting. Another strange thing is in the video, it appears to have a tiny black ball flying around the far left of the video (probe?), but the low clarity makes it difficult to make assumptions on it.

Was the sighting of this mother ship UFO over the capital city of Russian meant as a sign that aliens knew about NASA's planned moon bombing that would take place October 9th, 2009? For more information about the moon bombing, Google NASA, moon bombing, 2009. It would be foolish to assume that the two were not intermingled somehow. Perhaps this was a early warning that humans were overstepping their bounds and not only trespassing, but blowing up areas of the solar system that do not belong to humans alone. This would explain a lot more too. During the NASA moon bombing, which I watched over live NASA Internet, when the bomb impacted upon the surface of the moon, there was no explosion, no six-mile dust cloud that NASA scientists had predicted. Why you ask? Easy, the warhead was disarmed before it impacted the moons surface, but the real question is who disarmed it? The aliens...or NASA?

Chapter Twenty-Two: 2009 UFO Sighting by Fox News on Gilliland's ECETI Ranch

When Fox news heard about a man near Trout Lake, Washington who owns a ranch that has UFO sightings virtually every night, they just had to investigate it for themselves. James Gilliland owns a spiritual retreat center near Mt. Adams and he claims that UFOs have flown over his property every night for the last thirteen years.

47. Photo of UFO over Mt. Adams. (At http://scwbook. blogspot.com/).

Gilliland claims to have seen UFO ships extremely close up. He decided to integrate UFOs into his business and he often has people visit his property every night in hopes of catching sight of one or more UFOs. Gilliland says that the people that come to visit his ranch will often see unusual things flying above in the night sky.

Gilliland stated, "The ships coming here have come in every size, shape and color imaginable." He said that they have seen every glowing color imaginable flying in the night sky right above his ranch. He went on to say, "We have seen just about everything up here. We have even seen flying pyramids." Then he continued, "The ships we see up here are just so diverse, (referring to the different shapes and colors) that there is just no way it could be some back-engineered project or something like that."

When Mr. Gilliland said "back-engineered," what he was referring to was the government projects where the USAF will take actual captured UFOs and try to disassemble them so that they can figure out how and why they work as they do. My chapter on Area S4 in this book will shed more light on that subject.

Fox News stayed on Mr. Gilliland's ranch the night that they interviewed him and they actually were able to record video of the UFOs that they had heard so much about. They were able to record numerous glowing green ball like objects moving at incredibly high speeds through the night sky. In the video of the UFO that Fox News recorded, there can be seen a glowing light greenish ball of light flying past the stars. At different times, this UFO would pulse or blink, but always continued on its way. They revealed two of the recorded objects on their news for the public to see. In one video, the greenish glowing craft is moving to the left, passing many stars in the background. In the second video, a similar looking craft is moving to the right this time.

When Fox asked Mr. Gilliland about the UFOs seen over his property, he stated, "The universe is vast because if you take the two hundred billion suns, just in this little Milky Way Galaxy with planets revolving around it, and then you add to that the five hundred billion galaxies that are just like this one out there, its so vast and its inhabited, and there are beings that are thousands even billions of years ahead of us out there."

Mr. Gilliland is a very upfront and honest person when it comes to talking about the UFOs that fly over his ranch. He is even kind to the skeptics that say that UFOs don't exist and never have. Mr. Gilliland likes to invite the skeptics over to his ranch so that they may see the phenomenon flying above his ranch with their very own eyes. He understands that talking about UFOs and seeing them for yourself are two very different things. Many of the skeptics have agreed to come by and have a look for themselves.

When asked about who some of the skeptics were that came to the ranch, Mr. Gilliland states, "We have travel police on record. We have triple PHD Boeing engineers. Air Force base commanders, uh…the list goes on and on. Pilots and air traffic controllers and thousands of people have come here and seen the ships. We have had so much evidence that it's just a mountain of evidence that any reasonable mind would have to say that there is something fantastic going on here."

One eyewitness even stated, "The two objects were apparently round with a reddish, orange glow. Movements of the objects were erratic and discontinuous. They appeared to move independently, circling and changing places in relation to one another."

Mr. Gilliland said, "the government had an extreme role in keeping this information from coming out, and it goes back to even before Roswell. It goes back to the 1930s, but they have actually had interactions and they have actually connected with low level contacts and worked with them and gleaned technology from the ETs themselves, and also the crash sites and things of that nature. Fiber optics, microwaves, computer chips, all of that stuff came out of Roswell. It went to Bell labs then was disseminated from there on out."

Mr. Gilliland showed Fox News the satellite tracking data for the area. He showed them two different sets of tracking data from two different Internet sources. Both reports said that the sky would be clear that night and no satellites would be visible above his ranch area. He uses two websites to get the satellite information, just in case one may be wrong.

He says that, "When we see something, we can use this data to prove that its not anything that we have flying up there, but that it's some other object, an unknown."

Mr. Gilliland said, "It's such a massive universe. There's got to be life out there. If we are the best the creator has to offer, then it's a pretty sad situation."

In November of 2000, KATU TV in Portland, Oregon caught wind of the UFO sightings at Mr. Gilliland's ranch and interviewed him about it. They stated that, "Mt. Adams has become a hotbed for UFO believers."

In the KATU TV interview, Mr. Gilliland said, "We usually see them in the early afternoons towards up to three in the morning." Then he went on to say, "When it started happening, I actually thought I was losing it and going crazy." Then he went on to say something that was literally out of this world. He said that he had been taken up to a ship twice. He says that the aliens are humanoid and come from the Orion and Pleiades group of stars.

In the interview by KATU TV, one eyewitness only known as Megan stated, "I saw five ships total or flying objects."

Another eyewitness, Carolyn Gray stated, "I saw a golden globe against a blue sky."

One UFO investigator Spar Giedeman even insisted that around Mr. Gilliland, he could see a "pre-contact energy."

In the interview with the Portland TV news, Mr. Gilliland stated, "One of the main reasons they're coming here is because we live according to the universal principals that are necessary for a healthy society and environment, and that's the way they live, and we've met the protocols for the contact."

On a more skeptical note, in the same article by KATU TV, they note that Mr. Gilliland is known around Trout Lake as, "Cosmic Jim." One resident of Trout Lake, Brian Smith said, "I've never seen a UFO. I've talked to the guy all the time. He gets gas from us. He tries to tell us they're out there. I've never seen one."

Oddly enough Mr. Gilliland's ranch is not alone in the local area at seeing UFOs over his ranch. For several decades, the nearby Yakima Indian Reservation has also seen the strange

UFOs in the sky that they say comes and leaves from the nearby Adams Mountain.

Mr. Gilliland's personal website is really very well made and professional in appearance. It tells you information about such things as upcoming events at the ranch, Kunlun Nei Gung at ECETI, UFO Sighting map of the area, 10 years of UFO sightings at the ranch and so forth. The website has numerous videos of recording UFO orbs at his ranch. These orbs are not just above, but when watching some of the videos, the orbs will actually swoop down in front of the camera and then away. The close-ups of these strange lighted objects and the photo library of evidence that he has at this web site for anyone to view, clearly indicates that serious investigation by the US government is necessary. The website is at www.ECETI.org, go and check it out and if you get the courage, pick up your DVD camera and go have a visit for yourself to the ranch or the surrounding areas. Mr. Gilliland noted many times that the UFOs come out and return back into the mountainside. There are even photographs of these strange craft on his website on the side of Adams mountain, so you know the exact place on the mountain you need to visit.

It is not a coincidence that Mt. Adams has a large amount of UFO sightings every night. Rather, mountains and old volcano's are the hotbeds for UFO activity and sightings. Why you ask? Many believe that mountains are high enough that clouds and snow obscure the view of most people, so that the ships could enter and exit hidden tunnels that may lead to underground bases. Just try Googling the words, "UFO Sighting, mountain," or "UFO Sighting, volcano." The UFOs that people see near the mountains are often seen disappearing near its surface, instead of landing, so one can assume, everything has to go somewhere. Remember that they are not gods, but highly intelligent species whose technology far outweighs our own.

When Mr. Gilliland knows the crazy things people are saying behind his back and he states, "There's a lot of skeptics that say we've gone off the deep end. We just have overwhelming evidence, and we tell them that basically condemnation without investigation is the height of ignorance." Wise words that transcend to all areas, not just that of UFO research.

Mr. Gilliland knows that there are a lot of non-believers out there in the world. It seems that just the mere mention of the word UFO or aliens changes their entire view of you, as if you were some sort of lunatic. It's the same mentality of 400 years ago when the world insisted that Earth was the center of the universe. Yet one brave person who was quickly labeled a lunatic came forward to challenge this. His name was Galileo. His reward for the truth was life imprisonment in his home by the Catholic guards. What I am saying here is that to view humans on Earth as the most intelligent species in the whole universe would be like a goldfish in a park pond who may be the biggest in the pond and therefore believes it's the biggest compared to any fish in existence, without ever knowing 75% of the Earth is coved by water with creatures hundreds of thousands of times bigger than itself. Do you catch my drift on what I'm trying to say here? I'm saying don't jump to conclusions about those who report and research about UFO sightings, because one day, it may be you, and will you practice what you preach then? It's time to raise humanities level of awareness.

Chapter Twenty-Three: Spiral UFO seen hovering over Norway, 2009 as President Obama arrives for Nobel Peace Prize.

In the early morning hours, a mysterious giant spiral of light was seen dominating the dark sky over Norway. The giant spiral formation could be seen growing, and then getting smaller, then growing again in size. This spiral lasted for almost three minutes, with thousands of Norwegians recording it on video and in photographs. People flooded Norway's Meteorological Institute with phone calls asking for more information on the spiral light in the sky. This spiral was not just seen over a single city, but rather was seen from over five hundred miles away. Scientists were baffled at what the incredible spiral light could be with theories ranging from a misfired Russian missile, meteor, northern lights, black hole and finally a UFO. After the media spread the news of this spiral light across the world, a few days latter the Russians announced it was a misfired missile, launched from a submarine. This excuse however does not fit what other misfired missiles looked like in the air, and was done to try to cover up this UFO phenomenon that has spread worldwide. This sighting was not the only spiral in the sky ever recorded on video. There have been others of such sightings in different countries.

48. Photo of spiral UFO seen over Norway, 2009. (At http://scwbook.blogspot.com/).

The unusual about this sighting that caught my attention was that President Obama had flown in that day and a few days latter in Norway to receive the Nobel Peace Prize. The prize is awarded annually by the Norwegian Nobel Committee in accordance with the guidelines created by Alfred Nobel's will that began in 1901. The prize includes a medal, a personal diploma, and 10 million Swedish crowns ($1,500,000 US). This is an incredible award for a president to receive and perhaps the UFO spiral that was sighted that night, was there to congratulate Obama and to inspire him to keep on this road that Obama was just beginning. This fact of the spiral and Obama's visit to receive the award has not been widely publicized by the press. Instead they have separated the two events, refusing to consider that the two are linked. It's not like this is the first time a UFO was seen around President Obama. He has had two other UFO sightings recorded near him during his swearing in as president and as he was campaigning to be president.

The Norwegian Nobel Committee stated that Obama has already started a new peaceful way of working things out in international politics. The committee was impressed with Obama's methods of using dialogue and negotiations to resolve most international conflicts.

Perhaps then we can assume that the spiral was meant as a sign that Obama was being watched on all fronts, and not just by humans. Aliens having interest in Obama does not seem too surprising due to his insistence that words, not weapons should be used to resolve most conflicts.

Let's explore some of those possible explanations for the turquoise spiral that the news reported about. First off, the northern lights have never before in history been seen in Norway.

Secondly a meteor cannot fly in a spiral pattern within earth's atmosphere, although in the vacuum of space it may a possibility.

Thirdly, during the sighting, Russia flatly denied that it was a rogue rocket of theirs, but later changes their statement, saying it was one of their missiles. The Russians have such advanced missile and rocket systems compared to the USA that NASA often asks Russia to bring much needed supplies or crew members to the International Space Station (ISS), and even has the Russian rockets bring some crew back to Earth. Its unlikely they would risk such a display and possibly have a missile fall into a heavily populated area of Norway.

Fourth, it was a black hole. The center of the spiral did suddenly open into a pitch-black circle for 10-15 seconds, which was so black that no lights or stars could be seen behind or within it. This black hole effect is easily seen in many videos. A black hole is understandable and might even be possible since Norway is a few hundred miles from the CERN project in Geneva where the CERN's Large Hadron Collider (LHC) has become the world's highest energy particle accelerator, having accelerated its twin beams of protons to an energy of 1.18 TeV in the early morning hours of Nov. 30, 2009. (Note the only 1 TeV is required to create a mini black hole). Then the morning of the spiral, CERN announced that they completed a major test that morning. So the black hole theory cannot be eliminated so easily. One obvious thing that dismisses this theory is that there is no apparent connection between the black hole and Obama's visit to Norway.

Thousands if not millions of Norwegians' were eyewitnesses to this spiral that could be seen across Norway. It was an incredible spectacle as the white trail spiraled again and again around itself until the spiral was so incredibly big, had it been day, it could have blocked out the sun many times over. And by no means is this UFO spiral over Norway the only one to have ever been recorded on video.

On June of 2006, a UFO Spiral was seen over China along the ocean that could be seen for hundreds of miles away. This sighting was recorded not by one city, but by two different cities. Reporters went to the city of Da-Lein, not far from Beijing city. An eyewitness to the spiral stated this in Chinese, "I was the first person to see the flying object. It was moving in a spiral pattern. It was not flying straight."

Then flying toward an airport at a nearby city, flight 6358 flight attendants saw the same UFO spiral. Their unique vantage point of seeing the UFO from their airliner rather than the ground shows that the UFO spiral was not a trick of the sky, but an actual craft. More eyewitness on the ground said they got a really clear view of the spiral UFO from Da-Lein. They said the object was very high in the sky and was very bright with a long lit up tail. The spiral moved like a wheel rolling across the evening sky. It was moving from southeast to northwest. Some witnesses to the UFO believed that it was an airplane about to crash, causing the spiral, but it never fell to the ground. In China, many such UFO spirals have been recorded and seen by hundreds of thousands of people ever since the 1970s.

Another such event took place on August 27, 1987 at Shun-Hi Weather Station. At this station an employee saw a great spiral object that traveled across the sky at 8 PM. At seeing the spiral, she called the local airport radar tower, but there was nothing seen on the radar monitor. This eyewitness had many years experience of seeing my kinds of weather phenomenon, yet this event she had no explanation for. She stated in Chinese, "It was flying from West to East in a spiral formation. I saw it for about one or two minutes before it disappeared. It was about the size of a rice bowl with a long fire tail."

The spiral being actually a black hole seems to have some merit when you look at some of the scientists around the world that have said that the CERN project could in fact, accidently make black holes. Some scientists suggest that the LHC machine could reach collision energies at which gravity becomes a very strong force and small black holes may be produced in such collisions. Would the black hole first eat the LHC itself, then Geneva and then the world? By no means would this happen. Black holes with a mass around 1 TeV will not live long enough to pull anything in.

Stephen Hawkings stated, "they would be super-hot little objects that would dissipate all their energy very rapidly by emitting radiation and particles before they wink out of existence." Sure you ask, is there a standard model that scientist have hypothesized to form a black hole? Yes. New research suggests that gravity becomes more powerful at small distances because of the effects of extra dimensions used only by gravity. In this research, as the effective value of G grows bigger, the Planck mass drops, and the energy required to produce black holes can drop to 1 TeV. This is the same range the LHC reaches. In this way, the LHC (CERN) may turn out to be a "black hole assembly line," r an accelerator that makes large quantities of mini black holes that evaporate anywhere from a few seconds to a few minutes. Can you imagine the panic if the CERN project admitted to accidently making a few black holes? They would be instantly shut down, end of project and billions of dollars wasted. Scientific achievement always has its pluses and negatives, lets hope it's worth it. It could be used to create better technology and better energy sources. Yet the fact that Obama was in Norway to receive the award around the same time as the spiral, makes the theory of a black hole appear unlikely. Obama and the spiral over Norway seem too connected to disregard.

Overall, The fact that Obama was arriving in Norway the same time this spiral appeared, makes a firm foundation for believing that the two are linked. This UFO sighting is not the first seen near President Obama, but is actually the third, and surly will not be the last. Why would aliens be interested in Obama? Perhaps because he chooses diplomacy and words as his tools rather than force. Think of it this way, humans have urges that are both primitive (genetic) and advanced (thinking beyond our genetics). Perhaps the aliens recognize that Obama is such a man, recognizing in him a higher level of thinking that can push away the primitive (using force) urges that humans have within them. You decide which is the more advanced way of thinking, using force to make things happen, or using understanding and diplomacy? We are humans and our computers and society are advancing every day, but our attitudes and stubbornness has been around since the rise of mankind, and will continue to be around until the fall of mankind.

Conclusion

The information in this UFO book clearly identifies and weights the facts and possibilities that exist. NASA scientists themselves have reported to the press that the Shuttle Atlantis was being followed by several round silver metallic orbs (Shuttle Atlantis 2006) that they called UFOs, but they will no way believe that someone in the civilian sector could possibly identify a UFO. NASA has continually denied the existence of extraterrestrial life forms, yet again actual eyewitness reports of seeing UFOs include names of US presidents, US astronauts, military pilots, generals and many more people is such prestigious positions that we consider them to be heroes of our country. NASA says that the statements of UFO sightings are the opinions of astronauts that worked for NASA and are certainly not the opinions of NASA themselves. If NASA had it their way, they would not inform the world about their findings for another two hundred years. By that time, America would be in serious control of massive amounts of alien technology, both military and civilian, which would allow America to continue to be the most powerful country in the world.

America is clearly in control of a lot of alien technology at this very moment, so much so that they have long passed the testing stage and have moved into actual use of alien built crafts. These crafts vary in design greatly from one to the other. I have found on Goggle Earth three craft. Two saucer shaped craft and one triangle craft that are currently in Area S-4.

Remember that S-4 is (13.5 miles SW of Area 51) the place at Area-51 where they take and study alien technology in hopes of back engineering it. I have placed the photos of the craft in this book with the coordinates for finding them on Goggle Earth map. You see, its not enough for me to tell you what there is and isn't out there, but I have to be able to teach you how to use the same techniques that I used, in this way, you can use the techniques to confirm for yourselves what I have told you and what NASA has told the public. Please do not just take my word for it or theirs for that matter. Go out and explore for yourself. Your opinions and discoveries matter. I already told you in the chapters how to search and where to search, so if you trust your own eyes, you too will learn what the world governments have been hiding from the public for fifty years.

The existence of alien life forms and alien technology is obviously the greatest scientific discovery in human existence. My overall goal in this book and life is…to raise humanities level of awareness of the possibilities that exist around us. No small task mind you, but I believe I will achieve this goal.

I have accomplished in a few months of research what NASA says they still have not discovered in their fifty years, showing that alien buildings and ships are unmistakably visible in many NASA photographs. It is clear that if I could find the existence of alien life by using NASA photographs, then in the fifty years that NASA has existed, they have been either holding back the truth or looking for the wrong things. It is also possible that NASA itself was created as a cover-up for a more covert operation to help disguise their true mission, to gain alliances with alien cultures in hopes of gaining the upper hand with trade agreements and off world assignments for military personnel.

As a little boy I remember way back in first grade, how I admired and looked up to NASA for what they had achieved and what they must have seen beyond this world. I was in first grade when I first found the face on Mars in a huge NASA book, thirty-four years ago. I still have respect for what they have accomplished, yet I cannot condone their deceitful practices of hiding the truth from Joe Public.

If you want to learn more, then please pick up your digital cameras or DVD cameras and go to find some NASA photographs, enlarge them by 5X, then start scanning them on a flat computer screen while looking through your digital camera. Note, holding the camera at a slight angle (100-120 degrees) up will reveal hidden items that will first appear as ghostly clouds over the ground, but will then reveal to be structures. Digital begets digital, meaning that the digital photographs on a digital flat computer screen, seen through your digital camera will cause the once blurry 5X photo to re-digitalize and become focused again, sometimes gaining 100% focus. It may take a half hour or more to find your first structure, but once you gain in experience, it will get easier and easier till, like me, it takes mere minutes if I use a photograph from the Apollo Image Atlas at http://www.lpi.usra.edu/resources/apollo/. Then from there, go directly to the word 'Panoramic.' These are the easiest NASA images to enlarge and see structures with windows or doors. Sometimes the structures are even in the shape of an alien head or face. One building I found was in the shape of three aliens faces, each different making me believe it was a commemorative building of different species working together.

Try this, go to an Apollo mission photo area. Apollo 15,16,17 are all available. Apollo 15 has 1529 images. Apollo 16 has 1,435 images. Apollo 17 has 1581 images available for you. Then you must follow these simple steps:

1. Pick a photo, click on it. Then push your right button on your mouse to "save as," then choose to save it to your "desktop."

2. Once on desktop, click the picture using your right mouse button again, choose "open with," then choose "Windows Picture and Fax Viewer."

3. Once open, click on the magnifying glass + icon then click on the photo to make it larger. Enlarge it 3-5 times its original size (300-500%).

4. Once finished, start scanning over it with your digital camera. I found that my old Sony Cyber-shot with Full HD still images and 10.1 Mega Pixels gives a great view of what is really on the moon's surface.

5. Often, especially in the panoramas of Apollo Image Atlas, there are hidden structures, that will be revealed if you move your camera angle sweeping back and forth from a 0 degree angle to 30 degree angle, actually looking upwards at the screen. This reveals much more than you would believe, so the only way for you to understand is to try it and see for yourself. I'm not sure why this works, but it does. The first time you try this successfully, it will take your breath away.

This technique will work on most digital astronomical photographs, so now you can start exploring the universe in the comforts of your own home. Explore the moon, Mars, IO (my favorite) and all the other planets around. If you really feel adventurous and you are still not getting enough, look up some Hubble Telescope photographs and explore the cosmos, but mind you, this is only for those who already have experience at using the techniques I taught earlier. You will see things that defy explanation, learn more in a few hours than any astrophysics professor can teach you in a lifetime! In Hubble images, I found flat black structures, silver metallic structures and much more.

All I ask is that you take the information you have found, and reveal it to others over the Internet or in your daily life. Share your findings in emails, making photo movies using free Microsoft moviemaker, and post it at any of the hundreds of free video websites for others to view. Together we will pry they truth out of NASA. Don't be like NASA and Bogart all the good stuff.

I have revealed these things to you, doing what I as a teacher am suppose to do, to teach, in hopes of enlightening. In turn, perhaps one day, it will be the student that becomes the teacher.

For more truth about aliens and UFOs please visit:
The Disclosure Project at www.disclosureproject.org/

Made in the USA
Lexington, KY
11 August 2010